REDUCING THE FUEL CONSUMPTION AND GREENHOUSE GAS EMISSIONS OF MEDIUM- AND HEAVY-DUTY VEHICLES, PHASE TWO

FIRST REPORT

Committee on Assessment of Technologies and Approaches for Reducing the Fuel Consumption of Medium- and Heavy-Duty Vehicles, Phase Two

Board on Energy and Environmental Systems

Division on Engineering and Physical Sciences

Transportation Research Board

NATIONAL RESEARCH COUNCIL
OF THE NATIONAL ACADEMIES

THE NATIONAL ACADEMIES PRESS
Washington, D.C.
www.nap.edu

THE NATIONAL ACADEMIES PRESS • 500 Fifth Street, NW • Washington, DC 20001

NOTICE: The project that is the subject of this report was approved by the Governing Board of the National Research Council, whose members are drawn from the councils of the National Academy of Sciences, the National Academy of Engineering, and the Institute of Medicine. The members of the committee responsible for the report were chosen for their special competences and with regard for appropriate balance.

This study was supported by cooperative agreement DTNH22-12-H-00389 between the National Academy of Sciences and the U.S. Department of Transportation, National Highway Traffic Administration. Any opinions, findings, conclusions, or recommendations expressed in this publication are those of the author(s) and do not necessarily reflect the views of the organizations or agencies that provided support for the project.

International Standard Book Number-13: 978-0-309-30237-1
International Standard Book Number-10: 0-309-30237-4

Additional copies of this report are available from the National Academies Press, 500 Fifth Street, NW, Keck 360, Washington, DC 20001; (800) 624-6242 or (202) 334-3313; http://www.nap.edu.

Printed in the United States of America

THE NATIONAL ACADEMIES

Advisers to the Nation on Science, Engineering, and Medicine

The **National Academy of Sciences** is a private, nonprofit, self-perpetuating society of distinguished scholars engaged in scientific and engineering research, dedicated to the furtherance of science and technology and to their use for the general welfare. Upon the authority of the charter granted to it by the Congress in 1863, the Academy has a mandate that requires it to advise the federal government on scientific and technical matters. Dr. Ralph J. Cicerone is president of the National Academy of Sciences.

The **National Academy of Engineering** was established in 1964, under the charter of the National Academy of Sciences, as a parallel organization of outstanding engineers. It is autonomous in its administration and in the selection of its members, sharing with the National Academy of Sciences the responsibility for advising the federal government. The National Academy of Engineering also sponsors engineering programs aimed at meeting national needs, encourages education and research, and recognizes the superior achievements of engineers. Dr. C. D. Mote, Jr., is president of the National Academy of Engineering.

The **Institute of Medicine** was established in 1970 by the National Academy of Sciences to secure the services of eminent members of appropriate professions in the examination of policy matters pertaining to the health of the public. The Institute acts under the responsibility given to the National Academy of Sciences by its congressional charter to be an adviser to the federal government and, upon its own initiative, to identify issues of medical care, research, and education. Dr. Victor J. Dzau is president of the Institute of Medicine.

The **National Research Council** was organized by the National Academy of Sciences in 1916 to associate the broad community of science and technology with the Academy's purposes of furthering knowledge and advising the federal government. Functioning in accordance with general policies determined by the Academy, the Council has become the principal operating agency of both the National Academy of Sciences and the National Academy of Engineering in providing services to the government, the public, and the scientific and engineering communities. The Council is administered jointly by both Academies and the Institute of Medicine. Dr. Ralph J. Cicerone and Dr. C. D. Mote, Jr., are chair and vice chair, respectively, of the National Research Council.

www.national-academies.org

COMMITTEE ON ASSESSMENT OF TECHNOLOGIES AND APPROACHES FOR REDUCING THE FUEL CONSUMPTION OF MEDIUM- AND HEAVY-DUTY VEHICLES, PHASE TWO

ANDREW BROWN, JR., NAE,[1] Delphi Corporation, Troy, Michigan, *Chair*
INES AZEVEDO, Carnegie Mellon University, Pittsburgh, Pennsylvania
RODICA BARANESCU, NAE, University of Illinois at Chicago
THOMAS CACKETTE, California Air Resources Board (retired), Sacramento
NIGEL N. CLARK, West Virginia University, Morgantown
RONALD GRAVES, Oak Ridge National Laboratory, Knoxville, Tennessee
DANIEL HANCOCK, NAE, General Motors (retired), Indianapolis, Indiana
W. MICHAEL HANEMANN, NAS,[2] Arizona State University, Tempe
WINSTON HARRINGTON, Resources for the Future, Washington, D.C.
GARY MARCHANT, Arizona State University, Tempe
PAUL MENIG, Tech-I-M, Sherwood, Oregon
DAVID F. MERRION, Merrion Expert Consulting, Brighton, Michigan
AMELIA REGAN, University of California, Irvine
MIKE ROETH, North American Council for Freight Efficiency, Fort Wayne, Indiana
GARY W. ROGERS, Independent Consultant, Birmingham, Michigan
CHARLES K. SALTER, Independent Consultant, Chambersburg, Pennsylvania
CHRISTINE VUJOVICH, Cummins, Inc. (retired), Columbus, Indiana
JOHN WOODROOFFE, University of Michigan Transportation Research Institute, Ann Arbor
MARTIN ZIMMERMAN, University of Michigan, Ann Arbor

Staff

MARTIN OFFUTT, Responsible Staff Officer, Board on Energy and Environmental Systems
JAMES J. ZUCCHETTO, Director, Board on Energy and Environmental Systems
ALAN CRANE, Senior Scientist, Board on Energy and Environmental Systems
JOSEPH MORRIS, Senior Program Officer, Transportation Research Board
E. JONATHAN YANGER, Research Associate, Board on Energy and Environmental Systems
LaNITA JONES, Administrative Coordinator
DANA CAINES, Financial Manager, Board on Energy and Environmental Systems

[1] NAE, National Academy of Engineering.
[2] NAS, National Academy of Sciences.

BOARD ON ENERGY AND ENVIRONMENTAL SYSTEMS

ANDREW BROWN, JR., NAE,[1] Delphi Corporation, Troy, Michigan, *Chair*
DAVID T. ALLEN, University of Texas, Austin
WILLIAM F. BANHOLZER, NAE, The Dow Chemical Company, Midland, Michigan
WILLIAM F. BRINKMAN, NAS, Princeton University
WILLIAM CAVANAUGH III, Retired Chairman, Progress Energy, Raleigh, North Carolina
PAUL A DECOTIS, Long Island Power Authority, Albany, New York
CHRISTINE EHLIG-ECONOMIDES, NAE, Texas A&M University, College Station, Texas
SHERRI GOODMAN, CNA, Alexandria, Virginia
NARAIN HINGORANI, NAE, Consultant, Los Altos Hills, California
DEBBIE NIEMEIER, University of California, Davis
MARGO OGE, U.S. Environmental Protection Agency (retired), McLean, Virginia
MICHAEL OPPENHEIMER, Princeton University, New Jersey
JACKALYNE PFANNENSTIEL, Consultant, Piedmont, California
DAN REICHER, Stanford University, Palo Alto, California
BERNARD ROBERTSON, NAE, Daimler-Chrysler Corporation (retired), Bloomfield Hills, Michigan
GARY ROGERS, Independent Consultant, Birmingham, Michigan
ALISON SILVERSTEIN, Consultant, Pflugerville, Texas
MARK H. THIEMENS, NAS,[2] University of California, San Diego
RICHARD WHITE, Oppenheimer & Company, New York, New York
ADRIAN ZACCARIA, Bechtel Group (retired), Frederick, Maryland
MARY LOU ZOBACK, NAS, Stanford University, Palo Alto, California

Staff

JAMES J. ZUCCHETTO, Director
JOHN HOLMES, Associate Director and Senior Program Officer
DANA CAINES, Financial Associate
ALAN CRANE, Senior Scientist
ELIZABETH EULLER, Project Assistant
LaNITA JONES, Administrative Coordinator
MARTIN OFFUTT, Senior Program Officer
E. JONATHAN YANGER, Research Associate

---

[1] NAE, National Academy of Engineering.
[2] NAS, National Academy of Sciences.

TRANSPORTATION RESEARCH BOARD 2014 EXECUTIVE COMMITTEE[1]

[1]Membership as of March 2014.

Preface

The fuel consumption and greenhouse gas (GHG) emissions of medium- and heavy-duty vehicles (MHDVs) have become a focus of legislative and regulatory action in the past few years. Section 101 of the Energy Independence and Security Act of 2007 (EISA 2007), Pub. L No. 110-140 §101, mandated the U.S. Department of Transportation to promulgate fuel consumption standards for MHDVs for the first time. In addition, Section 108 of that same Act required the Secretary of Transportation to contract with the National Academy of Sciences to undertake a study on the technologies and costs for improving fuel consumption in MHDVs and prepare follow-on reports at 5-year intervals.

In response to the Secretary's request, the National Research Council (NRC) in 2010 completed *Technologies and Approaches to Reducing the Fuel Consumption of Medium- and Heavy-Duty Vehicles,* referred to henceforth as the Phase One Report. The Phase One Report provided a series of findings and recommendations on the following: the development of a fuel consumption program for MHDVs; metrics for measuring MHDV fuel consumption; availability and costs of various technologies for reducing fuel consumption; potential indirect effects and externalities associated with fuel consumption standards for MHDVs; alternatives for the scope, stringency, certification methods, and compliance approach for the standards; and a suggested demonstration program to validate innovative certification procedures and regulatory elements.

Thereafter, in 2011, the National Highway Traffic Safety Administration (NHTSA) and the U.S. Environmental Protection Agency issued the Phase I Rule on fuel consumption and GHG emissions of MHDVs.

This report comprises the first periodic, 5-year follow-on to the NRC's 2010 report. The NRC formed the Committee on Technologies and Approaches for Reducing the Fuel Consumption of Medium- and Heavy-Duty Vehicles, Phase Two, for this purpose.

In the present report, the committee seeks to advise NHTSA as it revises its regulatory regime for MHDVs that meets the two agencies' objectives of (1) reducing in-use emissions of carbon dioxide from MHDVs; (2) reducing in-use emissions of other GHGs from MHDVs; and (3) improving in-use efficiency of fuel use in MHDVs—by driving innovation, advancement, adoption, and in-use balance of technology through regulation. At the same time, the committee seeks to advise on pathways to accomplish this, subject to the following constraints: (a) holding life-cycle cost of technology change or technology addition to an acceptable level; (b) holding capital cost of acquiring required new technology to an acceptable level; (c) acknowledging the importance of employing a balance of energy resources that offers national security; (d) avoiding near-term, precipitous regulatory changes that are disruptive to commercial planning; (e) ensuring that the vehicles offered for sale remain suited to their intended purposes and meet user requirements; (f) ensuring that the process used to demonstrate compliance is accurate, efficient, and not excessively burdensome; and (g) not eroding control of criteria pollutants or unregulated species that may have health effects.

Objectives 1, 2, and 3 are not fully congruent when fuels having different carbon content are considered, and when GHGs other than carbon dioxide are considered. In particular, GHG and efficiency are decoupled when the fuel and engine technology changes. Objectives 1, 2 and 3 also require that any regulation must reflect real-world activity and performance of vehicles. Constraints (a) and (b) suggest that the regulation and standards may stop short of driving best available technology or certain technology pathways. However, (a) and (b) do not go so far as to suggest that new technology must offer a positive return on investment for the consumer through reduced fuel usage: Needs for efficiency and GHG reduction may reach beyond economic drivers for change. Constraints (c), (d), and (e) may dictate that a single standard may not be reasonable because a mix of fuels may be needed and because these different fuels may not be capable of meeting a common standard if the standard is set too ambitiously. Constraint (f) may be in conflict with the real-world benefit implications of the objectives. Constraints (c) and (d) imply that the regulations should not close current or anticipated technology pathways without adequate notice to manufacturers and suppliers.

The committee is grateful to all of the federal agencies, original equipment manufacturers, suppliers and their respective associations, and nongovernmental organizations whose staff contributed significantly of their time and efforts to this NRC study, either by giving presentations at committee meetings or by responding to committee requests for information.

This report has been reviewed in draft form by individuals chosen for their diverse perspectives and technical expertise, in accordance with procedures approved by the NRC's Report Review Committee. The purpose of this independent review is to provide candid and critical comments that will assist the institution in making its published report as sound as possible and to ensure that the report meets institutional standards for objectivity, evidence, and responsiveness to the study charge. The review comments and draft manuscript remain confidential to protect the integrity of the deliberative process. We wish to thank the following individuals for their review of this report:

Steve Berry, NAS, University of Chicago,
Daniel Blower, University of Michigan Transportation Research Institute,
Rebecca Brewster, American Transportation Research Institute,
Mike Camosy, Auto Research Center,
David Foster, University of Wisconsin (retired),
Art Fraas, Resources for the Future,
Steve Hanson, Pepsi-FritoLay,
Stephen Kratzke, NHTSA (retired),
Margo Oge, International Council on Clean Transportation,
Joseph Prahl, Case Western Reserve University,
Bernard Robertson, NAE, DaimlerChrysler,
Aymeric Rousseau, Argonne National Laboratory, and
James Spearot, Mountain Ridgeline Consulting, LLC.

Although the reviewers listed above have provided many constructive comments and suggestions, they were not asked to endorse the conclusions or recommendations nor did they see the final draft of the report before its release. The review of this report was overseen by Elisabeth M. Drake, NAE, Massachusetts Institute of Technology. Appointed by the NRC, she was responsible for making certain that an independent examination of this report was carried out in accordance with institutional procedures and that all review comments were carefully considered. Responsibility for the final content of this report rests entirely with the authoring committee and the institution.

Andrew Brown, Jr., *Chair*
Committee on Assessment of Technologies and Approaches for Reducing the Fuel Consumption of Medium- and Heavy-Duty Vehicles, Phase Two

Contents

SUMMARY 1

1 INTRODUCTION 7
Background, 7
Comparison of EPA and NHTSA MHDV Fuel Consumption Regulations to NRC Phase One Report Recommendations, 8
Metrics Used in the Regulation, 8
Classes of Vehicles to Regulate, 9
Certification Procedures, 9
Pilot Program and Evolution of the Regulatory Regime, 10
Trailer Regulations, 12
Testing, 12
Other Recommendations in the NRC Phase One Report That Were Not Addressed by the Agencies, 13
Market and Regulatory Background Factors, 15
Natural Gas, 15
Biofuels, 17
Electrification, 18
Life-Cycle Analysis of Fuels, 19
Automated/Connected Vehicles, 20
Green Logistics, 20
Background Regulatory Changes, 20
References, 21

2 POTENTIAL FOR TECHNOLOGICAL CHANGE IN COMMERCIAL VEHICLES TO IMPACT FUTURE NHTSA REGULATIONS 23
Overview of Medium- and Heavy-Duty Vehicles (Classes 2b Through 8), 23
Power Train Technologies, 25
Vehicle Technology, 27
References, 27

3 CERTIFICATION AND COMPLIANCE PROCEDURES USING GEM 29
Development of Greenhouse Gas Emissions Model, 30
Description of GEM, 32
Use of GEM for Over-the-Road Tractors, 33
Analysis of GEM, 35
User Interface, Order Entry, and GEM Utility, 35
User-Specified Data Input and Hard-Coded Features of GEM, 35

Fixed Values in the GEM Code, 37
Vehicle and Component Integration, 37
References, 38

4 BASELINE INFORMATION ON MHDV FLEET AND METHODOLOGY FOR COLLECTION 40
Introduction, 40
Why Do We Need a Baseline? NHTSA Should Have a Baseline in Order to Inform Its Rulemaking, 40
What Is a Baseline?, 40
Why a Baseline?, 41
Which Year Should the Baseline Capture?, 41
Which Data Should the Baseline Contain?, 41
Criteria for a Good Baseline Data Collection Process, 42
Comments on NHTSA, SwRI, and Frost & Sullivan Survey, 42
Comments on the CalHEAT Report for the California Energy Commission, 42
Findings and Recommendations, 43
References, 43
Annex 4A: Other Sources of Baseline Data in the Industry, 44
Annex 4B: Additional Ways to Obtain Information in the Future, 51

5 NATURAL GAS VEHICLES: IMPACTS AND REGULATORY FRAMEWORK 52
Summary of Supply and Demand Trends for Natural Gas Fuel, 52
Natural Gas Engines and Vehicles, 54
Technology, 54
Infrastructure, 60
Expected Growth in Natural Gas Vehicle Population, 62
Regulatory Framework for Natural Gas Engines and Trucks, 63
Greenhouse Gas Emission and Fuel Economy Standards for Engines, 63
NG Engines, 64
Emission and Fuel Economy Standards for Complete Trucks, 64
Findings and Recommendations, 65
References, 65

6 REVIEW OF OPTIONS TO REDUCE ENERGY USE OF TRAILERS 67
Background, 68
Current Tractor-Trailer Energy Balance, 68
Aerodynamics and Tire Rolling Resistance of the Tractor-Trailer, 68
Aerodynamics of the Combined Tractor-Trailer, 69
Tractor Aerodynamics, 69
Van Trailer Aerodynamics, 70
Tractor-Trailer Gap, 70
Tire Rolling Resistance, 70
Government Programs That Influence Tractor-Trailer Fuel Consumption, 72
SmartWay, 72
California Air Resources Board Regulation, 74
NHTSA and EPA Regulations, 74
Methods for Aerodynamic Performance Evaluation, 74
Current Use of Aerodynamic Devices and Low-Rolling-Resistance Tires, 76
Tractors, 76
Aerodynamic Devices on Van Trailers, 76
Market for Trailer Aerodynamic Devices, 77
Barriers to Increased Use of Trailer Aerodynamic Devices, 79
Tires, 80

Tire Pressure Systems, 82
Findings and Recommendations, 83
Trailers, 83
Tractors, 84
Tractors and Trailers, 84
Tires, 84
References, 84
Annex 6A: Questions Posed to Van Trailer Manufacturers to Gather Information for Table 6-5, 86
Annex 6B, 87

APPENDIXES

A Committee Biographical Information 91
B Statement of Task 97
C Committee Activities 98
D Acronyms and Abbreviations 100
E Glossary 102

Summary

BACKGROUND

Medium- and heavy-duty trucks, motor coaches, and transit buses—collectively, medium- and heavy-duty vehicles (MHDVs)[1]—are used in every sector of the economy. The purpose of these vehicles ranges from carrying passengers to moving goods. The fuel consumption and greenhouse gas (GHG) emissions of MHDVs have become a focus of legislative and regulatory action in the past few years. This report is a follow-on to the National Research Council's (NRC's) *Technologies and Approaches to Reducing the Fuel Consumption of Medium- and Heavy-Duty Vehicles* (NRC, 2010; henceforth called the Phase One Report), issued March 31, 2010. That report provided a series of findings and recommendations on the development of regulations for reducing fuel consumption of MHDVs.

On September 15, 2011, the National Highway Traffic Safety Administration (NHTSA) and the U.S. Environmental Protection Agency (EPA), hereinafter referred to as the Agencies, jointly published a Federal Register notice (76 Fed. Reg. 57105) finalizing rules to establish a comprehensive Heavy-Duty National Program to reduce GHG emissions and fuel consumption for on-road MHDVs (the Phase I Rule, also know as the Phase I Regulation).

Subsequently, NHTSA entered into a cooperative agreement with the NRC to issue a report by 2016 on technologies and approaches to reducing the fuel consumption of MHDVs with a view toward beginning work on a revision to the Phase I Rule. The committee has since learned that NHTSA and EPA have commenced work on a second round (Phase II) of fuel consumption and GHG emission standards for MHDVs. This first report by the committee provides guidance for the Phase II Rule, which is directed at the post-2018 time frame.[2] The committee's final report, to be issued in 2016, will cover a broader range of technologies and issues and will address the 2025-2030 time frame.

COMPARISON OF PHASE I RULE WITH THE PHASE ONE REPORT

In 2010, the NRC prepared a report in preparation for the Phase I regulation, proposing several recommendations. On the whole, NHTSA was quite responsive to the Phase One Report, in particular to the matter of taking the Report's advice on basing the standards on load-specific fuel consumption (LSFC)[3] and using modeling to determine compliance. Some recommendations were not adopted by NHTSA, and the committee encourages NHTSA and EPA to do so. In particular, the following aspects remain valid: the importance of data acquisition on the baseline MHDV stock, sales, and performance, without which conclusions on whether the goals of the regulation are met, will necessarily be very uncertain; the formation of an expert group to review vehicle simulations, in particular for the Greenhouse gas Emissions Model (GEM); implementation of driver training programs that may enable reduced fuel consumption; and incremental fuel efficiency gains that could be obtained by dieselization of Classes 2b through 7 vehicles. There were other recommendations that NHTSA did not act on, but the above seem to the committee to be the most compelling.

CERTIFICATION USING MODELING AND SIMULATION

From both technical and operational perspectives there is an objective to improve, and even optimize, efficiency metrics for trucks in each class. The existing NHTSA and

[1] More precisely, these vehicles include those classified as Class 2b through Class 8, which as a group range in gross (combined) vehicle weight from 8,500 to 80,000 pounds.

[2] This Summary contains the key recommendations from the committee's report.

[3] The precise metric for measuring fuel consumption, the LSFC is measured in gallons of fuel per payload ton per 100 miles. The lower the fuel consumption (FC) of the vehicle and the higher the payload the vehicle carries, the lower will be the LSFC.

EPA rules have already addressed the metrics to be employed for engine efficiency, vehicle efficiency, and the associated carbon dioxide (CO_2) emissions levels, for both gasoline and diesel. From a practical standpoint at the time of writing, the rules also drive change to provide economic benefit to the truck user. However, as truck efficiency regulation advances, there are trade-offs that must be addressed. Metrics of interest are fuel efficiency, GHGs, cost, criteria pollutants, and energy security. The primary trade-off is GHGs versus fuel efficiency, when several fuels and their associated technologies are considered.

Recommendation S.1: NHTSA, in consultation with EPA, should consider carefully the impact on related metrics when attempting to optimize for a single metric, or should otherwise establish a clearly articulated objective that weights, or places limits upon, relevant metrics. (Recommendation 3.1)

GEM is a relatively simple model focused on aerodynamics, rolling resistance, speed, weight, and idle control, and as such it is not capable of acknowledging efficiency and GHG emissions changes associated with engine and transmission design, the integration of advanced power trains, alternative fuels, hybrid and electric vehicles, and optimal component management. The weight reduction input in GEM is limited to a fixed set of technologies and parts. GEM upgrades are required to provide more realistic prediction of fuel use and GHG emissions, particularly as detailed in the next three recommendations.

Models should be capable of simulating real-world component behavior accurately and should not be oversimplified. GEM specifically does not allow for synergy between components, the operation or control of components in a most efficient way, or the engendering of efficiency through operation of a smaller component at higher relative load. GEM specifies the performance maps for major components such as the engine and transmission and does not credit the vehicle manufacturer with benefits of using a potentially superior engine or transmission.

Recommendation S.2: NHTSA should investigate allowing the original equipment manufacturer (OEM) to substitute OEM-specific models or code for the fixed models in the current GEM, including substituting a power pack (the engine, aftertreatment, transmission). These models, whether provided by OEMs or fixed in the code, should be configured to reflect real-world operation accurately. (Recommendation 3.7)

Calibration of equipment traceable to national standards for passenger car tires is in place, and self-calibration of equipment is in place for the grading of tires in Europe. Calibration of tire characterization equipment for U.S. regulations should be extended to cover all measurements of coefficients of rolling resistance used in GEM. For tire rolling resistance, there needs to be high confidence in the values inputted to GEM.

Recommendation S.3: A mechanism needs to be implemented for obtaining accurate tire rolling-resistance factors, including equipment calibration, and maintaining that information in a public database. This might be managed in the same way that tread wear, temperature, and traction data are displayed through the federal Uniform Tire Quality Grading system. (Recommendation 3.4)

GEM employs a limited set of cycles to challenge the simulated truck. These cycles do not include real-world road grade and neglect varying operating weights and aerodynamic yaw angles. Being speed-time based, these cycles also do not allow for the faster acceleration of more powerful trucks, or the longer time that might be taken by less powerful trucks to complete some real-world distance-based routes. This deficiency is not evident in the current model, where a hard-coded engine map, rather than a real OEM engine model, is used.

Recommendation S.4: The choice of test cycles and routes or schedules used in GEM needs to be readdressed thoroughly to avoid creating designs that are optimized for the test rather than for achieving real-world performance in the design process. (Recommendation 3.6)

FLEET DATA

A further issue of importance is the need to collect data on vehicles such as would permit regulators to evaluate the regulatory efficacy and improve both the accuracy of the Phase I Rule and any subsequent phases. NHTSA did not include a pilot phase when it promulgated the Phase I Rule in 2011. Thus there were no baseline data from even a few representative national fleets prior to the rulemaking, such as would have enabled comparison with post-rulemaking fuel efficiency. This would have also started to facilitate the comparison of real-world test data with compliance data. The committee nonetheless recognizes that NHTSA has begun the process of designing surveys and seeking the necessary Office of Management and Budget approvals to allow it to assemble a picture of the fleet characteristics, including the collection of R.L. Polk registration data and forecasts, which is appropriate for Class 2b to Class 8 vehicles.

Recommendation S.5: NHTSA should establish a repeatable, reliable data collection process as soon as possible. In addition to continuing data procurement with SwRI and R.L Polk, NHTSA should investigate outside sources—such as FTR, ACT Research, SmartWay, the North American Council for Freight Efficiency (NACFE), and the American Transportation Research Institute (ATRI)—to obtain a repeatable, reliable baseline as well as future data. These sources could use the R.L. Polk data and conduct clarifying, deeper interviews with truck and trailer builders, manufacturers, and fleets to elicit specific ongoing data on technologies procured and fuel consumption. (Recommendation 4.2)

NATURAL GAS

Natural gas accounts for about 25 percent of all U.S. energy use, yet only 0.1 percent is used in transportation, equivalent to about 0.5 billion gallons per year of petroleum fuel. However, in the short time since the release of the Phase One Report (NRC, 2010), natural gas has emerged as an economically attractive option for commercial vehicles. This has been driven by the rapid development of low-cost production of unconventional natural gas.

In order for medium and heavy trucks to use natural gas fuel rather than diesel, the most significant changes needed are the onboard fuel storage method and the means of introducing and igniting the fuel in the engine. Onboad fuel storage is by high pressure, effected by either compressed natural gas (CNG) cylinders (3,600 pounds per square inch is typical) or cryogenic containers filled with liquefied natural gas (LNG). For using natural gas in place of gasoline, the spark ignition engine carries over with modest changes, but fuel storage is still by one of the above two methods.

Natural gas engines are well developed, although improvements can be pursued in engine efficiency, maintenance costs, and onboard vehicle storage costs. Natural gas's inherent GHG benefit by virtue of its low carbon content (~28%) is partially negated by lower efficiency in currently available engines and the higher GHG impact of methane emissions. In addition, a natural gas leakage correction to GHG impact could negate the inherent tailpipe CO_2 advantage.

Recommendation S.6: NHTSA and EPA should develop a separate standard for natural gas vehicles as is presently the case for diesel- and gasoline-fueled vehicles. Factors the Agencies should consider in setting the standard include the maximum feasible ability of natural gas engines to achieve reductions in GHG emissions and fuel consumption, the uncertainties involved with the various alternatives, the impact of duty cycles on the ability to comply with the vehicle standards, the cost of the technology, and rapid growth of the market for natural gas engines and vehicles. This may require additional focused studies. (Recommendation 5.2)

Recommendation S.7: More studies and data are needed to determine the well-to-tank GHG emissions of natural gas vehicles, since current estimates vary significantly regarding quantification of emissions leakage of methane. EPA and NHTSA should assemble a best estimate of well-to-tank GHG emissions to be used as a context for developing future rulemakings. (Recommendation 5.1)

Due to the economics-driven rapid adoption of natural gas, there is urgency to develop an optimum solution in Phase II Rule standards for both GHG emissions and fuel consumption (as well as criteria emissions) that will accommodate this fuel without artificially disrupting prevailing commercial transportation business models. As a specific example, the GEM certification tools need to include natural gas engine maps to more accurately quantify the emissions and fuel economy of natural gas vehicles.

Recommendation S.8: To benefit fully from the GHG and petroleum displacement potential of natural gas, government and the private sector should support further technical improvements in engine efficiency and operating costs, reduction of storage costs, and emission controls (as is done for diesel engines). NHTSA and EPA should also evaluate the need for and benefits and costs of an in-use natural gas fuel specification for motor vehicle use. (Recommendation 5.4)

REGULATING TRAILERS

Aerodynamic Devices for Trailers

There are four regions of the tractor–van trailer combination truck that are amenable to aerodynamic design improvements, including the various tractor details, the tractor-trailer gap, the trailer underbody, and the trailer tail. Side skirts constitute 90 percent of devices sold to improve aerodynamics. Most side skirts provide a 5 percent fuel saving at 65 mph (3 percent at 55 mph) on EPA testing as part of SmartWay.[4] A recent study identified barriers to use of trailers that are more aerodynamically efficient (NACFE and Cascade Sierra Solutions, 2013), primarily related to understanding of cost-benefit data and lack of robust application information. Yet, early adopting carriers indicate there is a good return on investment from their use. Nonetheless, manufacturers of trailer side skirts report sales doubled between 2012 and 2013. The manufacturers also report installed prices of side skirts have declined by half.

A California regulation requires operators of van trailers to use aerodynamic devices to reduce the energy required to pull them. Observations made in California and Arizona showed a greater proportion of trailers with aerodynamic devices than did those observations made in Oregon, Texas, Michigan, Pennsylvania, and Maryland. Side skirts were overwhelmingly the predominant aerodynamic devices strategy. Other strategies (underbody fairings and rear fairings) were observed in just a few instances.

[4] SmartWay is a voluntary program administered by EPA that was formed in 2004 with the objective of improving efficiency and reducing fuel consumption and pollution from movement of freight across the supply chain. Currently it focuses mainly on over-the-road trucking.

Recommendation S.9: NHTSA, in coordination with EPA, should adopt a regulation requiring that all new 53 foot and longer dry van and refrigerated van trailers meet performance standards that will reduce their fuel consumption and CO_2 emissions. The lead time to implement this regulation should be evaluated independently from lead time requirements applicable to the next set of standards for new engines and tractors, because less time is needed to perform compliance testing and install aerodynamic devices on new trailers. The agencies should also collect real-world data on fleet use of aerodynamic trailers to help inform the regulation. (Recommendation 6.1)

The current SmartWay program and CARB regulation only address the most commonly used 53+ foot van trailer, which accounts for about 60 percent of the trailers that could potentially benefit from use of aerodynamic devices. Use of aerodynamic devices on other types of trailers, such as container/chassis and shorter vans including dual trailers ("pups"), could provide additional fuel savings of 4 to 9 percent per tractor-trailer, according to industry estimates. Fuel savings from use of side skirts have also been demonstrated on flatbed trailers. The cost-effectiveness of using aerodynamic devices on these additional categories of trailers depends on their annual mileage accumulation and average speed, among other considerations such as access to the trailer underbody, and needs further assessment and quantification.

Recommendation S.10: NHTSA, in coordination with EPA, should determine whether it would be practical and cost-effective to include along with the regulation of van trailers the regulation of other types of trailers such as pups, flatbeds, and container carriers, as doing so could substantially increase overall fuel savings. (Recommendation 6.2)

Various complete truck test procedures, such as the Society of Automotive Engineers (SAE) J1321 and coast-down procedure SAE J1263, used for determining the effectiveness of aerodynamic devices, are not sufficiently precise to discern small incremental changes. Alternatives such as wind tunnel testing and computational fluid dynamics (CFD) simulation should be evaluated because they provide fidelity and better precision. These methods can reduce the cost of development validation, may avoid building of multiple full-size prototypes, and can accelerate development time for the final product. Better test precision and repeatability may induce otherwise skeptical end users to adopt such technologies.

Recommendation S.11: NHTSA should evaluate the relative fidelities of the coast-down procedure and candidate powered procedures to define an optimum prescribed full-vehicle test procedure and process and should validate the improved procedure against real-world vehicle testing. Further, the Agencies should assess if adding yaw loads to the validation process provides significantly increased value to the drag coefficient (C_d) result. In addition, the Agencies should disseminate to end users updated test data and fuel savings of efficient trailers, aerodynamic devices, and tires, especially to those not participating in the SmartWay program. This should increase end user confidence in fuel savings and device reliability. (Recommendation 6.3)

Tires

Many new tractors and most new trailers are equipped with low-rolling-resistance tires that meet the SmartWay performance standard, and this is likely to increase due to regulatory requirements. However, 70 percent of new tires sold in 2012 for use on tractors and trailers were for replacement of existing tires, and only 42 percent of these are SmartWay verified. There is no assurance in the future that replacement tires will be as energy efficient as the original equipment tires they replace. Manufacturers have also introduced wide-base single tires (WBSTs), many of which feature lower rolling resistance than most dual-tire sets. WBSTs make up less than 10 percent of the commercial truck tire market, but their use is increasing as fleets strive to reduce fuel consumption and GHG emissions.

Recommendation S.12: NHTSA, in coordination with EPA, should further evaluate and quantify the rolling resistance of new tires, especially those sold as replacements. If additional cost-effective fuel savings can be achieved, NHTSA should adopt a regulation establishing a low-rolling-resistance performance standard for all new tires designed for tractor and trailer use. (Recommendation 6.5)

Precision in tire rolling resistance coefficient (C_{rr}) measurement is mandatory. Further, while the ISO28580 test procedure is given good grades by most in the industry, a robust machine cross-correlation does not exist for commercial vehicle tires in the United States. Carriers cannot depend on the comparability of C_{rr} for measurements from approximately 60 tire suppliers verified by SmartWay.

Recommendation S.13: NHTSA, supported by EPA, should expeditiously establish and validate the equipment and process for a tire industry machine alignment laboratory and mandate the use of that laboratory by each tire manufacturer seeking C_{rr} validation for any tires being offered as candidates in the GEM computation process, just as the C_{rr}'s of light-duty-vehicle tires were validated. (Recommendation 6.6)

OTHER APPROACHES TO REDUCING FUEL CONSUMPTION

The Phase I Rule had the effect of encouraging the adoption of technologies for reducing fuel consumption. Such reductions can be achieved by technological improvements to the vehicle as well as by improvements in operations, changes in behavior of drivers, and so forth. The Phase One Report considered other approaches (referred to, perhaps imprecisely, as nontechnical approaches) such as intelligent transportation systems; construction of lanes exclusively for trucks; congestion pricing; driver training; and intermodal operations (NRC, 2010, pp. 159 et seq.). Also considered were market-based instruments such as fuel taxes. Another viable approach would entail adjusting size and weight restrictions on trucks. For example, this might include greater use of vehicles that have favorable LSFC such as longer combination vehicles, which have greater freight capacity than the notional tractor-trailer, which can have a combined gross vehicle weight of 80,000 lb.[5] While some nonvehicle alternative approaches for reducing fuel consumption and greenhouse gas emissions may be beyond NHTSA's delegated authority, the agency can work with other agencies with appropriate authority as well as encourage private actors to consider such strategies to complement and support NHTSA's standards.

Recommendation S.14: NHTSA should consider additional strategies to encourage the adoption of measures that reduce fuel consumption by attempting to quantify the impacts of nontechnological factors on the costs and feasibility of future efforts to improve fuel consumption. (Recommendations 1.9 and 1.11)

REGULATORY PROCESSES

The committee also has made several observations about the regulatory process. Currently NHTSA and EPA standards consider fuel efficiency of the vehicle (in gallons per ton-mile) and tailpipe CO_2 emissions (in grams of CO_2 per ton-mile) that need to be achieved, on average, by the mix of vehicles sold each year by each manufacturer. Manufacturers are likely to achieve these vehicle standards using a variety of different energy fuels and technologies. Failure to consider the well-to-wheel emissions of each combination of fuel and vehicle technology may lead to regulations that do not achieve the anticipated energy and GHG emissions savings. Further, there is the possibility that regulations could produce incentives and behaviors that may result in unintended consequences that could be beneficial or detrimental, such as the water contamination that resulted from the addition of methyl tertiary butyl ether (MTBE) to gasoline as an oxygenate intended to reduce ground-level ozone.

Recommendation S.15: NHTSA, in coordination with EPA, should begin to consider the well-to-wheel, life-cycle energy consumption and greenhouse emissions associated with different vehicle and energy technologies to ensure that future rulemakings best accomplish their overall goals. (Recommendation 1.10)

Recommendation S.16: NHTSA should conduct an analysis, including methods such as expert surveys and scenario analysis or red teaming, as appropriate, to anticipate and analyze potential unintended consequences of its regulations and to determine whether additional actions are warranted to try to minimize such impacts. NHTSA should undertake this analysis concurrently with its next revision to its regulation. (Recommendation 1.4)

Regarding the potential for technological change in the MY2019-2022 time frame, the committee, in its investigations to date, has not identified any combustion or other engine technologies beyond those identified in the NRC (2010) Phase One Report that would provide significant further fuel consumption reduction during the Phase II Rule time frame. However, those technologies identified in the Phase One Report should be updated with current projections for fuel consumption reduction and adjusted for system interactions when used in combination with each other.

A further issue relates to the timing and rate at which technology enters the marketplace. In establishing the stringency of a Phase II Rule, careful evaluation of technology penetrations is necessary. While the NHTSA-stated intent of the Phase I Rule was to base targets on off-the-shelf technologies, the new baseline for setting a Phase II Rule will be drawn from more current vehicles and will include consideration of the different degrees of penetration attained by both off-the-shelf and future/advanced technologies by 2018.

Recommendation S.17: NHTSA's Phase II Rule should take the current and projected incremental fuel consumption reductions and penetration rates of the various technologies into careful consideration: These incremental reductions and penetration rates should be updated from those that were projected in the Phase I rulemaking. Furthermore, system interactions should be evaluated for the effect on the projected incremental reductions whenever combinations of technologies are considered. (Recommendation 2.2)

REFERENCES

National Highway Traffic Safety Administration (NHTSA). 2010. Factors and Considerations for Establishing a Fuel Efficiency Regulatory Program for Commercial Medium- and Heavy-Duty Vehicles. DOT HS 811 XXX. Available at http://www.nhtsa.gov/staticfiles/rulemaking/pdf/cafe/NHTSA_Study_Trucks.pdf.

[5] This follows from the observation that "weighing out" is better for fuel consumption than "cubing out," which refers to filling up the cargo area before reaching the combined gross vehicle weight limit.

National Research Council (NRC). 2010. Technologies and Approaches to Reducing the Fuel Consumption of Medium- and Heavy-Duty Vehicles. Washington, D.C.: The National Academies Press.

North American Council for Freight Efficiency (NACFE) and Cascade Sierra Solutions. 2013. Barriers to the Increased Adoption of Fuel Efficiency Technologies in the North American On-Road Freight Sector. July.

1

Introduction

BACKGROUND

The fuel consumption and greenhouse gas (GHG) emissions of medium- and heavy-duty vehicles (MHDVs) have become a focus of legislative and regulatory action in the past few years. Section 101 of the Energy Independence and Security Act of 2007 (EISA 2007), Pub. L. No. 110-140 §101, mandated the U.S. Department of Transportation (DOT) to promulgate fuel consumption standards for MHDVs for the first time. The statute requires DOT to provide 4 years of lead time between promulgation and enforcement of fuel consumption standards and also requires a period of 3 years of stability once the standards are in effect.

Section 108 of EISA also required the Secretary of Transportation to contract with the National Academy of Sciences (NAS) to undertake a study on the technologies and costs for improving fuel consumption in MHDVs. Within one year of the completion of the NAS study, the DOT was required to undertake its own study of the practicalities in promulgating fuel efficiency standards for MHDVs. Upon completion of that report, DOT was instructed to promulgate by rulemaking a fuel efficiency program for MHDVs that is "designed to achieve the maximum feasible improvement" in fuel consumption and to "adopt and implement appropriate test methods, measurement metrics, fuel economy standards, and compliance and enforcement protocols that are appropriate, cost-effective, and technologically feasible for commercial medium- and heavy-duty on-highway vehicles and work trucks." (49 U.S.C. § 32902(k)(2))

The present report is a follow-on to the National Research Council's (NRC's) *Technologies and Approaches to Reducing the Fuel Consumption of Medium- and Heavy-Duty Vehicles* (NRC, 2010; henceforth referred to as the "Phase One Report"). The NRC, the operating arm of the NAS, established the Committee to Assess Fuel Consumption Technologies for Medium- and Heavy-Duty Vehicles (henceforth the "NRC Phase One Committee"), which held its first meeting in December 2008 and continued with information gathering, deliberations, and report drafting before releasing its report on March 31, 2010. The Phase One Report provided a series of findings and recommendations on the development of a fuel consumption program for MHDVs; metrics for measuring MHDV fuel consumption; availability and costs of various technologies for reducing fuel consumption; potential indirect effects and externalities associated with fuel consumption standards for MHDVs; alternatives for the scope, stringency, certification methods, and compliance approach for the standards; and a suggested demonstration program to validate innovative certification procedures and regulatory elements.

Shortly after the release of the NRC report, President Barack Obama, on May 21, 2010, directed the National Highway Traffic Safety Administration (NHTSA), on behalf of DOT, to issue MHDV fuel consumption standards in close coordination with GHG emissions standards to be promulgated for the same vehicles by the U.S. Environmental Protection Agency (EPA). Given the connection between fuel consumption and GHG emissions, a coordinated approach to fuel consumption and GHG standards would reduce regulatory costs and burdens and minimize inconsistent regulatory requirements by allowing manufacturers to build one set of vehicles to comply with both sets of standards.

In October 2010, NHTSA released its report responding to the NRC report (NHTSA, 2010). The NHTSA analysis was generally consistent with the findings and recommendations of the NRC report, with some differences (see the next section), including issues (e.g., regulation of commercial trailers) that NHTSA deferred to future rulemakings.

On September 15, 2011, NHTSA and EPA, referred to hereinafter as "the Agencies," jointly published a Federal Register notice (76 Fed. Reg. 57105) finalizing rules to establish a comprehensive Heavy-Duty National Program to reduce GHG emissions and fuel consumption for on-road medium- and heavy-duty vehicles. NHTSA adopted final fuel consumption standards under its statutory authority provided by EISA, and EPA adopted carbon dioxide (CO_2)

emission standards under its Clean Air Act authority. (These are discussed in more detail in Chapter 5 of this report, on natural gas.) The coordinated rules were both tailored to the same three regulatory categories of heavy-duty vehicles: (1) combination tractors; (2) heavy-duty pickup trucks and vans; and (3) vocational vehicles. The EPA GHG emission standards commenced with model year 2014,[1] whereas NHTSA's fuel efficiency standards will be voluntary in model years 2014 and 2015 and become mandatory in model year 2016, in order to comply with EISA's 4-year lead-time requirement.

Following promulgation of the initial standards, NHTSA and EPA have commenced work on a second round (Phase II Rule) of fuel efficiency and GHG emission standards for MHDVs. The current NHTSA fuel consumption standards take effect in MY2016 and must remain stable for at least 3 years under the statute. New standards must provide 4 years' lead time. Assuming the Phase II fuel consumption regulation is promulgated in calendar year 2015, the earliest the new fuel consumption standards could go into effect is MY2020, owing to the 4-year lead time requirement.[2]

President Obama issued The President's Climate Action Plan in June 2013 (White House, 2013, p. 8), which states that the administration plans to work with stakeholders "to develop post-2018 fuel consumption standards for heavy-duty vehicles to further reduce fuel consumption through the application of advanced cost-effective technologies and continue efforts to improve the efficiency of moving goods across the United States."

The EISA anticipates that the NRC will update its report at 5-year intervals through 2025. Pursuant to that statutory timeline, NHTSA entered into a cooperative agreement with the NRC to issue a final report by 2016. The NRC formed the Committee on Technologies and Approaches for Reducing the Fuel Consumption of MHDVs (see Appendix A for member biographies) in January 2013. Subsequently, the cooperative agreement was modified (see Appendix B for the statement of task) to include a report to be issued in early 2014 that would inform a possible Phase II rulemaking such as that contemplated in the President's Climate Action Plan.

[1] For purposes of this report, the term "model year" will be synonymous with "calendar year," because unlike the light-duty vehicle sector, model years vary significantly among MHDV manufacturers, and so for the sake of simplicity and uniformity the calendar year is often used as the rough approximation for model year (MY).

[2] If NHTSA adopted its standards in mid-2015 it could start applying those standards to vehicles certified after that equivalent date in 2019, creating a split model year. The statement of task therefore refers to the possibility of the Phase II standards beginning in 2019. For purposes of its analysis here, however, the committee will assume that the Phase II standards will begin to apply to the entire 2020 model year.

COMPARISON OF EPA AND NHTSA MHDV FUEL CONSUMPTION REGULATIONS TO NRC PHASE ONE REPORT RECOMMENDATIONS

This section looks back at the NRC report *Technologies and Approaches to Reducing the Fuel Consumption of Medium- and Heavy-Duty Vehicles* (2010), specifically the impact it has had on NHTSA's and EPA's rulemakings.[3] In the preamble to the proposal of the Phase I Rule, the Agencies provided a response explaining their rationale for accepting or rejecting the NRC's recommendations.[4] In what follows, the committee provides its own views on the relationship between the Phase I Rule and the key findings and recommendations of thc NRC Phase One Report that are of continued relevance.

Metrics Used in the Regulation

The Phase One Report included the following recommendation:

> **Recommendation 2-1.** Any regulation of medium- and heavy-duty-vehicle fuel consumption should use load-specific fuel consumption (LSFC) as the metric and be based on using an average (or typical) payload based on national data representative of the classes and duty cycle of the vehicle. Standards might require different values of LSFC due to the various functions of the vehicle classes, e.g., buses, utility, line haul, pickup, and delivery. Regulators need to use a common procedure to develop baseline LSFC data for various applications, to determine if separate standards are required for different vehicles that have a common function. Any data reporting or labeling should state an LSFC value at specified tons of payload. (NRC, 2010)

The Agencies (EPA and NHTSA) followed the NRC Phase One Committee recommendation to base the fuel consumption standard on the vehicle work accomplished, such as load-specific fuel consumption (LSFC). Class 7 and Class 8 trucks and vocational trucks have been addressed specifically in this manner, and Class 2b pickups are handled effectively as mentioned below. The Agencies also gave considerable thought and study to selecting representative drive cycles so as to ensure the regulation would reduce GHG emissions.

A further consideration is the gross vehicle weight assumed in the GHG Emissions Model (GEM) simulation, which for Class 8 vehicles is based on a payload weight of 38,000 lb, an intermediate load value. The Agencies adopted payload values for the GEM simulation calculations that are representative of real-world truck use, instead of merely

[3] 75 Fed. Reg. 74152 to 74456 (Federal Register/Vol. 75, No. 229/ Tuesday, November 30, 2010/Proposed Rules).

[4] See also *Factors and Considerations for Establishing a Fuel Efficiency Regulatory Program for Commercial Medium- and Heavy-Duty Vehicles* (NHTSA, 2010).

using the maximum gross combination vehicle weight rating (GCVWR) for the vehicle weight class. This captures the situation that over half of trucks on the road are volume limited,[5] meaning the trailer is filled up with containers without reaching the weight limit. In such a case the combined tractor trailer is not at full GCVW of 80,000 lb, the maximum allowed weight for un-permitted interstate transit.[6] Appropriately, the Agencies indicated the need and intent to gather additional data on the weight/payload in actual service. They addressed the work-factor metric in Class 2b by accounting for the payload capability of these vehicles in the rule instead of setting a payload for evaluation, which overall addresses the Phase One Committee recommendations. The Agencies chose not to consider a metric for volume-limited freight, which might otherwise have been useful in the assessment of longer combination vehicles (LCVs). The NRC Phase One Report found that such vehicles "offer potential fuel savings for the trucking sector that rival the savings available from technology adoption for certain vehicle classes and/or types" (NRC, 2010, p. 176). Payload and its relationship to LSFC remain important considerations.

Recommendation 1.1: NHTSA should evaluate the load-specific fuel consumption (LSFC) at more than one payload to ensure there is not an undesirable acute sensitivity to payload by a particular truck power train and to reflect the fact that some states allow vehicles to operate with gross combination vehicle weight ratings well in excess of the values adopted for the simulation.

Classes of Vehicles to Regulate

The Phase One Report included the following finding:

> **Finding 8-1.** While it may seem expedient to focus initially on those classes of vehicles with the largest fuel consumption (i.e., Class 8, Class 6, and Class 2b, which together account for approximately 90 percent of fuel consumption of MHDVs), the committee believes that selectively regulating only certain vehicle classes would lead to very serious unintended consequences and would compromise the intent of the regulation. Within vehicle classes, there may be certain subclasses of vehicles (e.g., fire trucks) that could be exempt from the regulation without creating market distortions. (NRC, 2010)

The Agencies agreed with the NRC that regulating all MHDV classes at the outset of the regulation was important. As noted in the Phase One Report, if NHTSA were to regulate only Classes 2b, 6, and 8, this would encompass 90 percent of the fuel used by all medium- and heavy-duty vehicles. The Phase One Committee was quick to note, however, that uneven policy application may cause disruptions in the marketplace and create the potential for reclassifying various classes of vehicles, as has been done in light-duty vehicles (LDVs). Other unintended consequences might result, such as changes in market behavior to avoid higher prices due to regulation (e.g., if Class 2b is regulated but not Class 3, then buyers might buy more of the larger Class 3 trucks because they would become less expensive relative to 2b trucks). In view of these considerations, the committee believes regulating all MHDVs should remain the "agencies' objective." As the regulatory framework becomes more defined and comprehensive, additional effort needs to be applied to avoid unintended consequences, as addressed in the Phase One Report.

Certification Procedures

The Agencies adopted the general recommendation of using simulation to handle the wide range of vehicle configurations and equipment and drive cycles, while building on existing protocols of engine testing for criteria emissions. The use of simulation for the vehicle, with a separate engine test, generally followed the NRC Phase One Report recommendation to certify entire vehicles. The Phase One Report included the following recommendation:

> **Recommendation 8-4.** Simulation modeling should be used with component test data and additional tested inputs from power train tests, which could lower the cost and administrative burden yet achieve the needed accuracy of results. This is similar to the approach taken in Japan, but with the important clarification that the program would represent all of the parameters of the vehicle (power train, aerodynamics, and tires) and relate fuel consumption to the vehicle task. (NRC, 2010)

The Agencies developed a relatively measured regulation in 2011 (EPA and NHTSA, 2011a) in that the fuel efficiency targets are modestly challenging for some vehicle classes, and the certification process builds largely on current methods. The exception where extensive engineering was required was the development of the GEM for Classes 2b-8 vehicle compliance (see Chapter 3 for more discussion of GEM). GEM, a MATLAB/Simulink-based model, uses the same physical principles as many other existing vehicle simulation models to derive governing equations that describe driveline components, engine, and vehicle. These equations are then integrated in time to calculate transient speed and torque (EPA and NHTSA, 2011b, p. 4-2). The development and benchmarking of GEM are found in EPA reports (EPA, 2011; EPA and NTHSA, 2011b). The Agencies reduced the engineering challenge by simplifying the model, excluding hybrid powertrains and several widely used component technologies (e.g., automatic transmissions).

---

[5] Federal Register 57158 states that "These payload values represent a heavily loaded trailer, but not maximum GVWR, since as described above the majority of tractors 'cube-out' rather than 'weigh-out.'"

[6] GCVW of more than twice this weight is possible with special permits on certain roadways.

To capture fully the fuel consumption benefits of technologies in future regulatory phases, more engineering will be needed (and in fact is under way). The committee notes, however, that the selected drive cycles do not include external effects such as road grade or cross-winds (i.e., yaw angle), which are particularly significant for Class 7 and Class 8 vehicles. The simulations will thus not fully reflect the benefits of certain types of technologies for reducing fuel consumption. The NRC Phase One Committee had noted that road grade variations, for example, are absent from practically all widely used test cycles. A further example is adaptive cruise control, a subsystem actuated by radar systems to set a desired speed and offering the option of maintaining a set following interval from a vehicle directly in front. The longitudinal control this technology provides offers co-benefits for fuel consumption. These systems have full systems control. Further, as noted by the agencies, many of the vehicle specifications (transmission and final drive ratio) are left to the routine specification process (EPA and NHTSA, 2011a, p. 57158). The GEM simulation is based on a few user input parameters, including rolling resistance, aerodynamic drag coefficient, and vehicle weight reductions (EPA and NHTSA, 2011b, p. 4-10). Hence the question of expected change in performance is not fully answered and neither is the question of whether the Phase I Rule will have a favorable impact.

A lingering concern in GEM is the inability of manufacturers of Class 8 vehicles to take into account the engine choice, actual engine efficiency, and integrated power train design optimization, items that can provide as much benefit as some aerodynamic features.[7] Also, the cooling system is neglected: It may be unfavorably affected by efficient aerodynamics yet not accounted for at the engine. The cooling system can represent 5 percent of the vehicle power demand during some operations.[8]

Both GEM and current test cycles are time based and may not accurately reflect fuel consumption to accomplish a mission. Improved productivity may not be recognized, yet it saves fuel because commercial vehicles will run until the mission (work, distance, etc.) is completed.

Finding: The current certification procedures rely on computer simulations that have only a few unbound variables that can be user-specified. Further, GEM specifically does not allow for synergy between components, the operation of components in a most efficient way, or the engendering of efficiency through operation of a smaller component at higher relative load. Vehicle designs that are optimized for the conditions of the simulation may not be optimized in the real-world operation.

Recommendation 1.2: NHTSA should conduct a real-world evaluation to validate the simulated fuel consumption outputs in light of the input data used. The evaluation should include a sensitivity analysis on key parameters, such as gross combination vehicle weight, to judge whether the variation in these parameters leads to a source of error in the simulation. NHTSA will need to test a reasonable number and variety of vehicles to further refine and validate the Greenhouse Gas Emissions Model (GEM) simulations.

Pilot Program and Evolution of the Regulatory Regime

The NRC Phase One Report included the following recommendation:

> **Recommendation 8-6.** NHTSA should conduct a pilot program to "test drive" the certification process and validate the regulatory instrument proof of concept. It should have these elements:
>
> - Gain experience with certification testing, data gathering, compiling, and reporting. There needs to be a concerted effort to determine the accuracy and repeatability of all the test methods and simulation strategies that will be used with any proposed regulatory standards and a willingness to fix issues that are found.
> - Gather data on fuel consumption from several representative fleets of vehicles. This should continue to provide a real-world check on the effectiveness of the regulatory design on the fuel consumption of trucking fleets in various parts of the marketplace and in various regions of the country. (NRC, 2010)

It appears that the administration's schedule for issuing a rule quickly was a key factor in not conducting a pilot program. It is recognized that the entire NHTSA regulation was on a mandatory fast track, as requested by President Obama (White House, 2010). The committee compliments the Agencies on getting a Phase I Rule in place quickly, to promote fuel savings as soon as feasible.

The recommendation that NHTSA conduct a pilot program had two broad purposes: first, the agency would gain experience with certification testing, data gathering, compiling, and reporting. The trial period was envisaged as a means for developing and refining the regulatory processes before the official start date of the program. Second, the pilot program would include gathering data on fuel consumption from several representative fleets of commercial trucks (e.g., long-haul, delivery vans, specialty vehicles, and large pickups). These data would provide a real world check on the effectiveness of the regulatory design on the fuel consumption of trucking fleets in various parts of the marketplace and in various regions of the country (NRC, 2010, p. 188).

[7] David Kayes, Daimler Trucks North America, "Lessons Learned from FE/GHG Phase 1 Regulations, and Ways to Incorporate the Most Likely Future Technologies into FE/GHG Phase 2 and 3 Regulations," Presentation to the committee, March 20, 2013.

[8] Nigel Clark, West Virginia University, Morgantown, "Engine Models and Maps for Truck Certification," Presentation to the committee, June 20, 2013.

The Agencies, however, declined to undertake a pilot program.[9] Data gathering and comparing the performance of vehicles specified via the Phase I Rulemaking process versus current methods of specifying trucks for customers (using OEM specification tools) could nonetheless have begun in 2011 and been continued until now. Data gathering should be ongoing. At least some kind of demonstration programs could have been done, perhaps even with simulations.

Omissions that were due to the absence of a demonstration program include the following:

1. The lack of baseline data from a few representative national fleets prior to the rulemaking, such as would enable comparison with post-rulemaking (after 2014) fuel efficiency. This would have also started to facilitate the comparison of real-world test data with compliance data. The committee nonetheless recognizes that NHTSA has begun the process of designing surveys and seeking the necessary approvals from the Office of Management and Budget[10] to allow it to assemble a picture of the fleet characteristics.[11]
2. Early assessment of the process, accuracy, and repeatability of both tire rolling resistance measurements and aerodynamic drag measurements—in particular of vehicles in Classes 7 and 8. These measurements rely on less proven methodologies than the engine fuel efficiency measurements, which rely on established emissions certifications procedures in a well-controlled test cell. The GEM model requires the insertion of drag coefficient and tire rolling resistance variables as two of the very few parameters in the model over which the manufacturer has control. It is important that all communities have confidence in accurate determination of these variables; otherwise, there may be a perception that GEM predictions and binning could be impacted.[12] (GEM is discussed in detail in Chapter 3.)

The MHDV regulatory regime has had a short history relative to other fuel economy regulatory programs. There has not been the opportunity to benefit from numerous cycles of learning, development of regulatory measures, data acquisition, demonstrations, research and development (R&D), and modeling and simulation such as might be incorporated in periodic revisions to the rule. Many of these processes, such as R&D and models development, can take several years to reach fruition.

Finding: NHTSA can expect to benefit from insights and learning from technological advances and stakeholder dialogue.

Recommendation 1.3: NHTSA should allow the process of revising its regulations to be informed by the research and development cycle; advances in model development; and data collection, including its ongoing effort to develop surveys of the current fleet.

Fleet Data

A further issue of importance is the need for data such as would permit regulators to evaluate regulatory efficacy. While the rejected Recommendation 8-6, quoted earlier, from the NRC Phase One Report assumed the "gathering of data from fleets" would occur prior to regulation, the NRC's notion remains highly valid and is still required to improve the accuracy of the original promulgation program and, most certainly, the next phases. Even though the current program is structured as an incremental approach, both sides of the increment (with and without the program change) need to be auditable to validate the declared improvement.

There has been no action to restart the Vehicle Inventory and Use Survey (VIUS) or any similar survey. Using VIUS, researchers at Argonne National Laboratory found that during idle, Class 8 sleeper cabs use 7 percent of their fuel (Gaines, 2006; Capps, 2008). The data provided by such a survey would also be very useful for safety analysis, freight planning, and transport system analysis. This is discussed further in Chapter 4.

Unintended Consequences

Interventions into complex systems inevitably produce unintended consequences. The American sociologist Robert Merton (1936) recognized that purposeful actions to try to change a system will often produce unintended effects that can be positive or negative. Some unintended consequences can be anticipated, while others cannot. An example of a beneficial unintended consequence is when cities started installing light-emitting diode (LED) lighting to save energy, accidents were also reduced because the lamps burnt out less frequently. An example of a detrimental unintended consequence is the water contamination that resulted from the addition of methyl tertiary butyl ether to gasoline as an oxygenate intended to reduce ground-level ozone. Sound public policy involves attempting to anticipate and reduce, when feasible and appropriate, the negative unintended consequences of policy interventions. Methods such as scenario

[9] 75 Fed. Reg. 74354.

[10] The Office of Management and Budget's (OMB's) Office of Information and Regulatory Affairs is a statutory office created to administer the Paperwork Reduction Act.

[11] See 77 Fed. Reg. 75257.

[12] Binning is a method employed in model development to represent the real-world characteristics of a vehicle in a stylized, discretized manner. It involves the creation of a predefined set of notional categories into which real-world vehicles are sorted for purposes of carrying out the simulation. For example, GEM utilizes five bins to represent the aerodynamic characteristics of various vehicle configurations (EPA and NHTSA, 2011b, p. 2-46).

analysis or "red teaming" can be used to formally investigate potential unintended consequences (Lempert, 2007).

Fuel consumption regulations, in purposely trying to change product characteristics and mixes, could produce incentives and behaviors that may result in unintended consequences, either beneficial or detrimental (see, for example, Yun, 1997; NRC, 2001; and Harrington and McConnell, 2003). Some analysts have noted that original equipment manufacturers (OEMs) responded to the Corporate Average Fuel Economy (CAFE) standards by producing vehicles that counted as trucks for regulatory purposes (NRC, 2001, p. 10; Harrington and McConnell, 2003, p. 11).

Of note, two heavy-duty gasoline engine manufacturers (Ford[13] and GM[14]) said that the Phase I Regulations are considerably more difficult to achieve for gasoline engines than they are for diesel engines in vehicle classes where both engine types are available (notably Classes 2b and 3). Both manufacturers have indicated that marginalization or elimination of gasoline engines from this segment is a possible future outcome based on present forecasts, and this feedback should be carefully considered when setting Phase II Regulations applicable to this segment. The Agencies may wish to consider whether such consequences are likely and, if so, to what extent they will be detrimental to the long-run health of the industry and the goals of reduced fuel consumption and GHG reduction, and if such second-order impacts can or should be mitigated.[15]

Finding: The avoidance of unintended consequences needs to continue to be an essential consideration during development of the MHDV regulations.

Recommendation 1.4: NHTSA should conduct an analysis, including methods such as expert surveys and scenario analysis or red teaming, as appropriate, to anticipate and analyze potential unintended consequences of its regulations and to determine whether additional actions are warranted to try to minimize such impacts. NHTSA should undertake this analysis concurrently with its next revision to its regulation.

[13] Ken McAlinden, Ford Motor Company, "Heavy Duty GHG from a Full-Line Manufacturer's Perspective." Presentation to the committee, June 20, 2013.

[14] Mark A. Allen and Barbara Kiss, "General Motors Comments: NAS Panel on Heavy-Duty GHG/CAFE Discussion," Presentation to the committee, July 31, 2013.

[15] The EPA faced a similar issue in its 2000 rulemaking on Tier 2 emission standards for light-duty vehicles. The Agency was concerned that its regulations would have unintended differentiated effects on diesel versus gasoline vehicles. EPA identified this potential concern in its rulemaking and proactively took measures to prevent any adverse unintended consequences by modifying the compliance schedule for diesel vehicles (EPA, 2000, pp. 6739-40).

Trailer Regulations

The committee believes the Agencies were prudent in not establishing regulations for trailers in their Phase I Rule,[16] given the additional time needed to develop and promulgate such regulations. In its report, the Phase One Committee found that "trailers, which present an important opportunity for fuel consumption reduction, can benefit from improvements in aerodynamics and tires." Noting the synergies between tractors and trailers, that earlier committee recommended that "separate regulation of trailer manufacturers will be necessary to promote more fuel-efficient trailers, including integration of the trailer design with the tractor for improved aerodynamic performance, lower tare weight, and a requirement for low-rolling-resistance tires" (NRC, 2010, p. 189). Harmonization of tractor and dry van and refrigerated trailer aerodynamic features was not possible for 2014, but awareness would have been raised by requiring minimum semitrailer performance, even in select categories.

Finding: The omission of trailer regulations has led to suboptimal regulatory constructs when considering the combined tractor-trailer. The culture change in the tractor-van trailer fleet technical community has progressed but has done so absent clear signals on the cost-effectiveness of integrating trailers into the total vehicle. Separate regulation of trailers for fuel efficiency will have the beneficial effect of beginning integration of trailer design with the tractor for improved aerodynamic performance, lower tare weight, and a requirement for low-rolling-resistance tires.

Testing

Aerodynamic Test Method

The Phase One Report included the following recommendation:

> **Recommendation 5-1.** Regulators should require that aerodynamic features be evaluated on a wind-averaged basis that takes into account the effects of yaw. Tractor and trailer manufacturers should be required to certify their drag coefficient results using a common industry standard. (NRC, 2010)

The Agencies understandably did not try to develop or implement a standard way of assessing aerodynamic drag coefficients of Classes 7 and 8 tractors in the Phase I Rule. Instead, they chose a reference method—an enhanced coastdown procedure—but at the same time included a process for manufacturers to calibrate results from their own test methods to said reference procedure.[17] This eased the administrative and test burden on the industry for this initial Phase. The

[16] 75 Fed. Reg. 75354; 76 Fed. Reg. 57111.

[17] 76 Fed. Reg. 57148-57149.

Agencies announced plans to perform further assessments for future phases.

The Phase I Rule adopted test procedures that may not consider yaw angle; however, it was noted that "the Agencies are adopting provisions which allow manufacturers to generate credits reflecting performance of technologies which improve the aerodynamic performance in crosswind conditions."[18] Apparent wind yaw angles are certainly important in characterizing aerodynamic efficiency: They have a significant influence on the magnitude of aerodynamic drag. Some designs are likely to be more "yaw resistant" than others, so that a simple determination of the drag coefficient at zero yaw angle may not promote the best design in practice. The binning of data does not truly provide error margins, and misclassification relative to the real benefit can and will occur. The complexity of aerodynamic issues may exceed the ability of even a demonstration program. Findings in this area have been delayed by not having such a program.

The committee believes there are sufficient wind tunnel facilities in North America (including scale model wind tunnels with moving ground and yaw capability) to facilitate measurements of drag at varying yaw angles. Not accounting for the performance of aerodynamic drag in relation to apparent wind yaw angle keeps devices that perform better in yaw conditions from being given proper credit. These issues potentially exist with aerodynamic measurements and therefore are required to improve both the accuracy of the current program and (most certainly) that of the next phases.

Recommendation 1.5: NHTSA should create a dedicated program focusing on aerodynamics of the regulated categories of vehicles that would allow these factors to be more accurately considered in the overall fuel consumption reduction of commercial vehicles of different classes. This will entail a program of experiments and dedicated instrumentation.

Component Testing

The Phase One Report included the following recommendation:

> **Recommendation 8-5.** Congress should appropriate money for and NHTSA should implement as soon as possible a major engineering contract that would analyze several actual vehicles covering several applications and develop an approach to component testing and related data collection in conjunction with vehicle simulation modeling to arrive at LSFC data for these vehicles. The actual vehicles should also be tested by appropriate full-scale test procedures to confirm the actual LSFC values and the reductions measured with fuel consumption reduction technologies in order to validate the evaluation method. (NRC, 2010)

[18] 76 Fed. Reg. 57149.

NHTSA has sponsored a project at Southwest Research Institute (SwRI) that began in early 2013.[19] Based on the SwRI representative's status report on June 21, 2013,[20] the committee concludes NHTSA is addressing Recommendation 8-5 of the Phase One Report.[21] The committee is highly supportive of these efforts to quantify fuel consumption benefits.

Finding: In vehicle simulation, there is uncertainty about the effectiveness of particular aerodynamic devices, as well as other components like accessories, auxiliary power units, driveline parts, and tires. Some standardized test protocols are needed. This concern is elaborated in Chapters 3 and 6 as it relates to aerodynamic devices and in Chapter 6 as it relates to tires.

Other Recommendations in the NRC Phase One Report That Were Not Addressed by the Agencies

The committee believes some further points and recommendations from the Phase One Report are meritorious and still worthy of consideration. One issue was establishing the point of compliance or point of regulation. The Agencies have, in fact, partially implemented the recommendation on regulating the final stage manufacturer, but they retained a separate engine regulation based on a generic engine. The latter relies on simulation to streamline the certification process and accommodate the wide range of whole-vehicle configurations.

The points of regulation in the Phase I Rule are as follows:[22]

- Class 2b pickups and vans are regulated at the final vehicle builder.
- Vocational trucks are regulated at both the engine manufacturer and the chassis builder.[23]
- Classes 7 and 8 are regulated at both the engine manufacturer stage and the final-stage manufacturer.

As described on pages 190-194 of the Phase One Report:

[19] Thomas Reinhart, Southwest Research Institute, "Phase Two MD/HD Vehicle Fuel Efficiency Technology Study," Presentation to the committee, June 21, 2013.

[20] Ibid.

[21] Recommendation 8-5 called for "a major engineering contract that would analyze several actual vehicles covering several applications and develop an approach to component testing and related data collection in conjunction with vehicle simulation modeling to arrive at LSFC data for these vehicles."

[22] 76 Fed. Reg. 57110.

[23] Regulating vocational trucks, such as dump trucks, refuse trucks, vacuum trucks, and so forth, poses unique challenges with respect to the high specification variation needed in order for a vehicle to perform its function and the added challenge that follows from the complexity of the bodies on these trucks, which can make up to 80 percent of the end-use vehicle content.

> **Finding 8-8:** A certification test method must be highly accurate, repeatable, and identical to the in-use compliance tests as is the case with current regulation of light-duty vehicles tested on a chassis dynamometer, and for heavy-duty engine emissions standards tested on engine dynamometers.

Further, Finding 8-9 from the Phase One Report says that "to account for the fuel consumption benefits of hybrid power trains and transmission technology, the present engine-only tests for emissions certification will need to be augmented with other power train components added to the engine test cell, either as real hardware or as simulated components." This led to Recommendation 8-4 in the Phase One Report, which says, in part, "Simulation modeling should be used with component test data and additional tested inputs from power train tests." The need to account for the close interaction of the engine with other components/subsystems, such as the aftertreatment subsystem and the transmission, is greater now that hybrids and automated mechanical transmissions and aftertreatment algorithms have been improved. Therefore there is a need to use validated models of integrated power train components or actual power pack data operated over real-world drive cycles in the certification process.

Recommendation 1.6: NHTSA should consider the option of power train (power pack) testing and certification and of using, in GEM, verified models of the actual power train (power pack) shipped with the vehicle at the time of manufacture.

Recommendation 1.7: NHTSA, in coordination with EPA, should work toward requiring quantified baselines and improvements, achieved via recognized test protocols, especially for the "Vehicle Power Demands" parameters of the actual vehicle's consumption of fuel process.

Further, NRC included in its Phase One Report the following recommendation for a study of dieselization of Class 2b through Class 7 vehicles:

> **Recommendation 4-2.** Because the potential for fuel consumption reduction through dieselization of Class 2b to 7 vehicles is high, the U.S. Department of Transportation/National Highway Traffic Safety Administration (NHTSA) should conduct a study of Class 2b to 7 vehicles regarding gasoline versus diesel engines considering the incremental fuel consumption reduction of diesels, the price of diesel versus gasoline engines in 2010-2011, especially considering the high cost of diesel emission control systems, and the diesel advantage in durability, with a focus on the costs and benefits of the dieselization of this fleet of vehicles.

Diesel engines present an opportunity for incremental fuel efficiency gains and, for some vehicles, may have the advantage of better durability. The Phase One Report also noted the high cost of diesel emission control systems and the price of diesel versus gasoline engines prevalent in 2010-2011 as factors that would prevent a move to a more fuel-efficient diesel power train. Since the Phase One Report was issued, natural gas engines have also become an option in some fleet applications. It remains relevant to determine the emission control strategy for NO_x and CO_2 emissions for Class 2b through Class 7 vehicles that would deliver the greatest reduction in fuel consumption and GHG emissions overall for this fleet, taking into account the cost of compliance.

Also in the Phase One Report was Recommendation 4-4:

> **Recommendation 4-4.** NHSTA should support the formation of an expert working group charged with evaluating available computer simulation tools for predicting fuel consumption reduction in medium- and heavy-duty vehicles and developing standards for further use and integration of these simulation tools.

While NHTSA did not take such actions, EPA requested expert inputs, conducted an external peer review (EPA, 2011), and compared simulation packages in the course of developing GEM. As a result, the agencies have not fully addressed this recommendation.

A further recommendation in the earlier report, Recommendation 5-2, reads as follows:

> **Recommendation 5-2.** There are numerous variables that contribute to the range of results of test programs. An industry standard (SAE) protocol for measuring and reporting the coefficient of rolling resistance is recommended to aid consumer selection, similar to that proposed for passenger car tires.

Although not specifically mentioned in the Federal Register notice promulgating the rule (EPA and NHTSA, 2011a), the International Organization for Standardization (ISO) published a standard practice in 2009 (ISO, 2009). This practice should be considered for inclusion in the future rulemaking.

The Phase I Rule had the effect of encouraging the adoption of technologies for reducing fuel consumption.[24] Such reductions can be achieved through technological improvements to the vehicle as well as by improvements in operations, changes in behavior, and so forth. The Phase One Report considered nontechnology approaches such as intelligent transportation systems; construction of exclusive truck lanes; congestion pricing; driver training; and intermodal operations (NRC, 2010, pp. 159 et seq.). Also considered were market-based instruments such as fuel taxes.

[24] The precise metric for measuring fuel consumption, the load-specific fuel consumption (LSFC) is measured in gallons of fuel per payload ton per 100 miles. The lower the fuel consumption (FC) of the vehicle and the higher the payload the vehicle carries, the lower the LSFC.

Another viable approach would entail adjusting size and weight restrictions on trucks. For example, this might include greater use of vehicles that have favorable load-specific fuel consumption—for example, LCVs, which have greater freight capacity than the notional tractor-trailer, which has a GCVW of 80,000 lb.[25] Egress for LCVs from the network of roads and highways they are constrained to operate within might become an issue as the number of such vehicles increases. This must not be undertaken at the expense of safety and must operate with appropriate permitting. Lastly, Recommendation 7-3 of NRC (2010) called for training drivers to reduce fuel consumption. Training requirements for commercial driving licenses could be influenced by the U.S. Department of Transportation and NHTSA.

Recommendation 1.8: Recommendations 4-2, 4-4, 5-2, and 7-3 from the 2010 National Research Council report *Technologies and Approaches to Reducing the Fuel Consumption of Medium- and Heavy-Duty Vehicles* should be given further consideration in future regulations.

Finding: A number of strategies are available for reducing MHDV fuel consumption that do not involve changes to the engine or vehicle. While some nonvehicle alternative approaches for reducing fuel consumption and GHG may be beyond NHTSA's delegated authority, the agency can and should work with other agencies with appropriate authority as well as to encourage private actors to consider such strategies to complement and support NHTSA's standards.

Recommendation 1.9: NHTSA should consider additional strategies to encourage the adoption of measures that reduce fuel consumption. It should work together with EPA, the Federal Highway Administration, and the U.S. Department of Energy's (DOE's) Office of Energy Efficiency and Renewable Energy, as applicable, to realize such fuel savings in the real world and to create knowledge and incentives to capture the benefits of approaches other than regulating the vehicle, such as by changes to fleet operations and logistics.

MARKET AND REGULATORY BACKGROUND FACTORS

This section briefly summarizes market and other regulatory background factors that are relevant to MHDV fuel consumption and GHG emissions that have significantly changed since the Phase One Report was completed. The Phase One Report reviewed the technologies that could practicably be considered for near-term and long-term fuel consumption and CO_2 reduction, and in addition discussed a significant number of technology and market trends. For instance, there was already a move from diesel to gasoline direct injection technology at the middle of the MHDV range, as noted in the Phase One Report (NRC, 2010, p. 64). Likewise a shift to more fuel efficient, smaller displacement, greater power-density diesel engines was then becoming apparent and was expected to motivate continued downsizing, as with passenger cars. An analysis by Frost & Sullivan, a consultancy, indicates 15 liter (15 L) engines will continue dominating the Class 8 engine market through 2018 but then are expected to lose market share to 11 L to 12 L and 12 L to 14 L engines.[26] While the consequences of these moves are reduced fuel consumption and reduced CO_2 emissions, they may also have implications for the market as a whole and may influence factors such as supply chain and fuel choice.

Given that the previous Phase One Report was so comprehensive, only a few additional market and other background regulatory factors are identified in the current report as potentially affecting the efficacy of a future regulation. Some of these factors might counteract or complicate attempts to achieve improvements in fuel consumption or reductions in GHG emissions through the anticipated Phase II Regulations of NHTSA and EPA, while other factors might help to achieve those improvements and thus help to make the regulation more feasible. In addressing these factors, the committee focuses primarily on those that are likely to have a significant effect from 2020 to 2025, which is the likely implementation period of the Phase II regulations.

Of course the state of the economy will have a significant impact on MHDV vehicle-miles traveled (VMT), fuel consumption, and GHG emissions, particularly as macroeconomic trends affect growth and activity in the construction and manufacturing industry sectors. But in addition to the general economic health condition of the nation, the other potentially relevant factors discussed here include (1) the emergence of natural gas as a significant transportation fuel; (2) the role of biofuels; (3) the growing interest in the United States in dimethyl ether (DME) as a fuel; (4) the viability of electrification of the vehicles; (5) the development of automated and/or connected vehicles; (6) the implementation of green logistics; and (7) background regulatory developments.

Natural Gas

The natural-gas-fueled engine, using either liquid natural gas (LNG) or compressed natural gas (CNG), is not a new technology. Natural gas engines were produced as early as

[25] This follows from the observation that "weighing out" is better for load-specific fuel consumption than "cubing out," with the latter referring to filling up the cargo area before reaching the GCVW limit.

[26] Sandeep Kar, Frost & Sullivan, "Strategic Outlook of North American Medium and Heavy Commercial Truck Powertrain Market Megatrends and Industry Focus Indicate a Cleaner and Smarter Freight Movement Environment," Presentation to the committee, March 21, 2013.

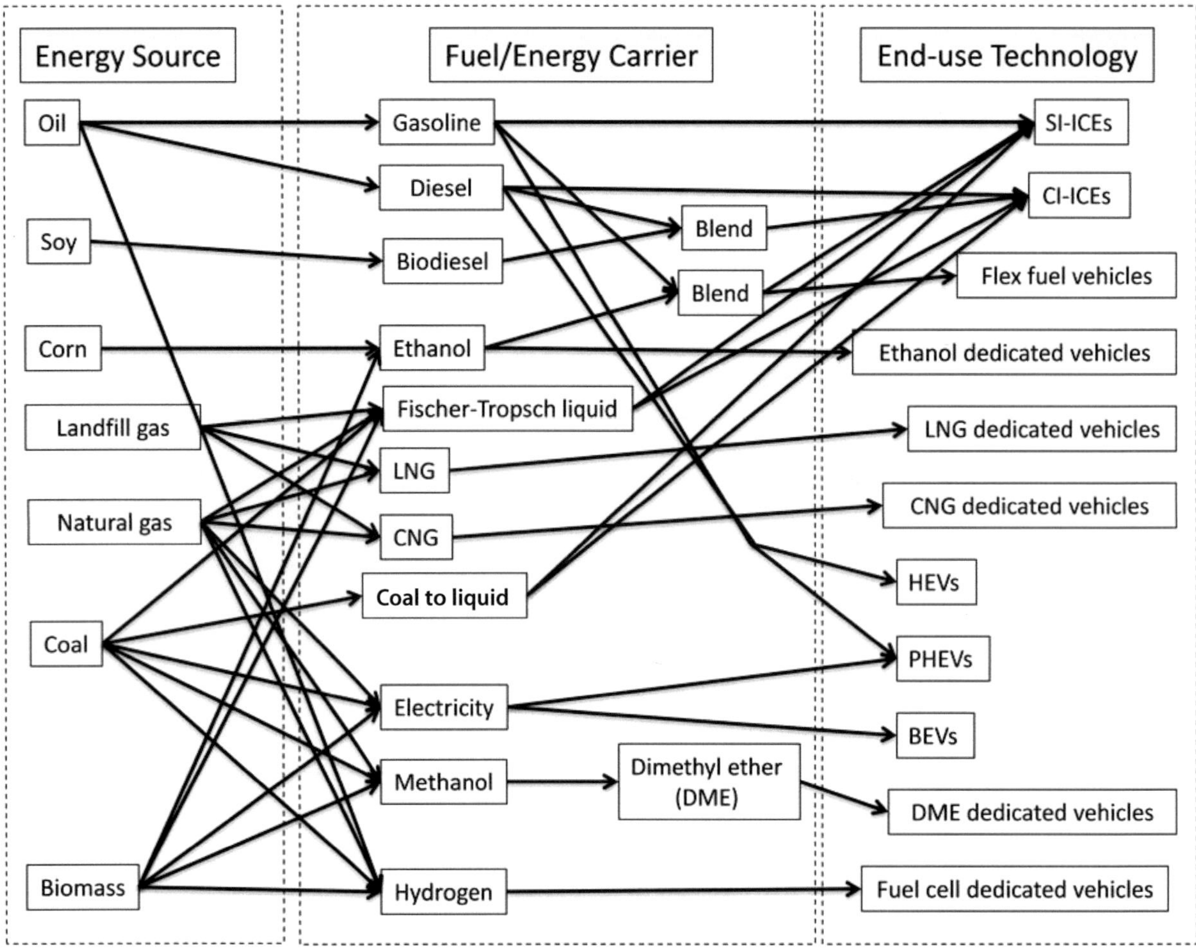

FIGURE 1-1 Illustrative pathway for vehicle fuels production and use.

1860 and now power about 120,000 vehicles on U.S. roads.[27] The application of natural gas for MHDVs has been more recent, however, and earliest uses were for transit buses and municipal vehicles. Over the past two decades, the natural gas engine has served as a niche technology in the MHDV market, present in mostly urban refuse haulers and transit bus applications. Natural gas is often referred to as a "bridge fuel," since it is a way to bridge the diesel-fuel dominance of the MHDV market to the next non-petroleum-based fuel—yet to be identified to the point of having a broad consensus. Common production pathways and uses for natural gas and other current and future MHDV fuels are illustrated in Figure 1-1.

The MHDV natural gas market developed slowly before circa 2010. Purchasers other than municipal fleets, which are subsidized by the government, had difficulty justifying the higher purchase price of the vehicle despite the lower cost of natural gas compared to diesel fuel. Furthermore, the cost of constructing fueling stations across the country ranges between $600,000 to over $1,000,000 per station for compressed natural gas and nearly twice that per liquefied natural gas station.[28]

Municipal vehicles, which run routes during the day and are centrally garaged at night, can be readily refueled at the garage, making them good applications for this niche technology.

In recent years, the gap between natural gas and diesel fuel prices has dramatically widened.[29] Moreover, advancements in technology have enabled manufacturers to develop

[27] http://www.ngvamerica.org/media_ctr/fact_ngv.html.

[28] TIAX (undated), "US and Canadian Natural Gas Vehicle Market Analysis—CNG Infrastructure: Final Report." Prepared for American Natural Gas Association. Available at http://anga.us/media/content/F7D3861D-9ADE-7964-0C27B6F29D0A662B/files/11_1803_anga_module5_cng_dd10.pdf; and TIAX (undated), "Gas Vehicle Market Analysis—LNG Infrastructure: Final Report," Prepared for American Natural Gas Association. Available at http://anga.us/media/content/F7D3861D-9ADE-7964-0C27B6F29D0A662B/files/LNG%20Infrastructure.pdf.

[29] ACT Research, 2012, "The Future of Natural Gas Engines in Heavy Duty Trucks: The Diesel of Tomorrow?" August 10.

more natural gas engine options and attendant vehicle technologies to achieve reliability and durability similar to that of the diesel. Together these circumstances make natural gas a viable choice for future commercial over-the-road fleets.

A variety of natural gas engines suited to a wide range of MHDV applications will be available by 2015. As more OEMs are introducing natural gas options to their product line, the share of CNG/LNG MHDVs continues to grow. ACT Research predicts[30] that the natural gas market share of MHDV truck and bus (includes municipal and refuse) could be as high as 36 percent by 2020. For these predictions to play out, the CNG/LNG infrastructure must be expanded. While there has been a significant increase in the number of natural gas fueling stations over the past years, the infrastructure is still nascent and will require large investments to provide enough stations to prevent disruption in routes and travel times for longer-haul trucks.

Another consideration in the future use of natural gas in the MHDV market is the rapid growth and output of hydraulic fracturing ("fracking") in natural gas drilling. Fracking has greatly increased the supply and availability of natural gas while reducing its cost. EPA and some states are now exploring more rigorous regulation of fracking operations. Regulations are one of several factors that could significantly increase the cost or reduce the availability of natural gas. This would reduce the incentive to move toward natural gas fuels and technologies in the MHDV sector.

Affordable fuel prices and a growing infrastructure all bode well for the future of natural gas in MHDVs. However, if the price of fuel continues to be favorable vis-à-vis diesel, the transportation sector will have to compete with other sectors (e.g., electricity and heating) for domestic natural gas. (The exporting of natural gas could affect prices as well.) Predicting how this might affect the MHDV market is difficult. Analysts predict that as the economy improves, the price of natural gas will increase (AEO, 2013) but so will the price of petroleum-based fuels.

Another important issue raised by fuels such as natural gas is, on the one hand. the distinction between vehicle fuel consumption and GHG and, on the other, the life-cycle analysis of the fuel consumption and GHG using natural gas as a fuel (well to wheels). This issue is addressed in Chapter 5 of this report, which discusses the role natural-gas-fueled MHDVs will play in the reduction of fuel use and CO_2 emissions in the future.

Biofuels

The current state of biofuel research, development, and production suggests that the biofuels produced in abundance over the next decade will likely be blends containing ethanol, gasoline, or biodiesel. In its 2013 *Energy Outlook*, the DOE's Energy Information Administration (EIA) forecasts that the consumption of next-generation biofuels (including pyrolysis oils, biomass-derived Fischer-Tropsch liquids, and fuels derived from renewable feedstocks) by the transportation sector will increase to about 0.4 million barrels per day (BPD) from 2011 to 2040. This compares with 1.6 million BPD of diesel during the same period. Given this presence of biofuels, the future fuel consumption and CO_2 reduction regulations for MHD trucks must take into consideration the effects of biofuels on the implementation of the future standards.

Ethanol

Ethanol has been used as a blend in gasoline engines for over three decades. Several federal regulations and programs have facilitated the use of ethanol as an oxygenate in the fuel to reduce air pollution (EPACT 2005 and EISA 2007). GHG emissions for E10 are 12 to 19 percent lower than those for pure gasoline (Argonne National Laboratory, 1999) at equal engine efficiency. Ethanol has the added benefit of reducing the U.S. dependence on petroleum, since it is made from plant materials, or "biomass." In 2001, the production of ethanol as a share of gasoline volume was only 1 percent. By 2011, the share rose to 10 percent (EIA, 2012). This is largely due to the first Renewable Fuel Standard (RFS) program, which was enacted in 2006 as a part of the Energy Policy Act of 2005. As a result of EISA 2007, the Renewable Fuels Standard "RFS2" mandated renewable fuel consumption of 36 billion gallons (35 billion of ethanol equivalent and 1 billion of biodiesel) by 2022.

Although higher blends of ethanol are approved as a transportation fuel by EPA (E15 and E85), the majority of vehicles in the United States use E10. Higher blends can produce fewer GHG, but the higher blends usually exhibit less "tank mileage" (miles per gallon), because of the inherent lower energy content (i.e., enthalpy) of ethanol. Every 10 percent of ethanol in the fuel reduces fuel economy by approximately 3.5 percent (Knoll et al., 2009). Further, distribution infrastructure becomes more difficult at higher blends. Ethanol is a solvent, so its chemistry is prone to dissolving the hydrocarbon residue and water that are often found in the pipeline, which can render the transported fuel out of specification, especially if tanks and pipes are not properly cleaned before switching products. In some cases where other blends of ethanol are desirable, filling station pumps are modified to blend pure gasoline with E85 to produce the new blend. Note that in the mid-1980s heavy-duty diesel engines were developed and demonstrated using ethanol fuel with an ignition assist.

Biodiesel

Studies by EPA and others indicate that the fuel consumption of B5, the most commonly used biodiesel, is about

[30] Ibid.

2 percent worse than that for conventional diesel.[31] However, vehicle CO_2 emissions for B20 can be 15 percent less than that for diesel. The B20 pumps are available at a growing number of outlets throughout the United States.

In 2001 biodiesel production was 9 million gallons. By 2011, it was nearly 100 times higher, at 967 million gallons. While this growth is significant, it represents only 1 percent of the total diesel production by volume. Consumption of biodiesel in 2011 was 878 million gallons (EIA, 2012). Similarly, RFS and EISA 2007 (RFS2) require consumption of 1 billion gallons biomass-based diesel. Tax credits and incentives through the RFS2 have had a positive influence on the production and consumption of biodiesel. Soybeans make up 57 percent of the biodiesel feedstock. Thus, droughts such as that the United States experienced in 2012 can cause the price of biodiesel to vacillate markedly, giving users little reason to purchase the fuel.

The use of biofuels is well established in the United States. The growth in production and consumption still relies in a large part on incentives and tax credits. Non-food-derived cellulosic feedstock is another consideration in the growth of these biofuels, but large-scale production and consumption is years away (NREL, 2012). A further fuel not yet in widespread use is so-called renewable diesel fuel, which is bio-oil refined to remove oxygen and which resembles petrol-derived fuels.

Fischer-Tropsch

Other alternative fuels, known as Fischer-Tropsch (FT) or gas-to-liquid (GTL) fuels, are available in the market but are currently produced in very modest volumes: only about 200,000 barrels a day, which is equal to less than 1 percent of global diesel demand a day (NYT, 2012). These fuels are produced via the FT chemical process, using natural gas, coal, or biomass as feedstock. FT fuels are interesting because they reduce dependency on crude oil and, depending on the feedstock used, may reduce the CO_2 footprint as compared with petroleum-based fuels. The resultant fuel from the FT process is a high-quality fuel (NRC, 2009). Hydrocarbon, NO_x, and particulate emissions all improve compared to diesel fuel when FT fuels are used (DOT-FHA, 2013).

These benefits notwithstanding, FT fuels are expensive to produce. Capital costs, the reliability of cost-effective feedstock, and the logistics of sourcing and transporting feedstock are all considerable. Analysts believe that FT fuels will be cost-effective only when natural gas and oil prices are out of balance. As long as natural gas and oil price differentials remain relatively aligned, the large investment in FT technology will be unsustainable (NYT, 2012).

[31] Petroleum diesel blended with 5 percent biodiesel.

Dimethyl Ether

Dimethyl ether (DME) may show promise as an alternative fuel. Synthesized from methanol, it can be produced from biomass, natural gas, or coal. Volvo Powertrain NA, the engine manufacturer for and supplier to Volvo and Mack truck brands, has announced it will produce engines operating on DME in 2015.

DME can have lower CO_2 emissions than conventional diesel on a well-to-wheels basis, particularly if the feedstock is biomass. The clean-burning characteristics of DME result in virtually no soot production, making a particulate filter unnecessary. Its thermal efficiency and performance are comparable to those of diesel. According to the International DME Association, DME typically sells at a premium to energy value (i.e., costs more for the same enthalpy). DME is liquefied at 50 pounds per square inch (psi) (or 345 kilopascal (kPa)), so its use requires similar tankage to propane. DME is expected to have the same selling price as a diesel gallon equivalent.[32] As with most alternative fuels, developing engine and vehicle modifications and the distribution infrastructure for the fuel are the most obvious obstacles to widespread use of DME in the near term.

DME currently has minimal transportation applications in the United States. It would be prudent, however, for the Agencies to consider its role in future transportation segments, given its growing popularity in Sweden and Japan and its forecasted presence in the United States.

Electrification

The electrification of the light-duty fleet appears to be finally achieving traction after many years of false starts and slow progress,[33] raising the potential for electric or hybrid medium- and heavy-duty vehicles to reduce CO_2 emissions and fuel consumption. There are a number of technology alternatives for incorporating electrification into the MHDV fleet, including (1) hybrid-electric vehicles (HEVs); (2) electrified accessories; (3) fully electric power trains; (4) electrified power take-off (PTO); (5) plug-in hybrid-electric vehicles (PHEVs); (6) external power to electric power train for zero emission vehicle (ZEV) corridors; and (7) alternative fuel/hybrid combinations.[34] In addition, there are so-called hotel load requirements to allow the driver of a Class 8 sleeper tractor to sleep in or otherwise occupy the

[32] Anthony Greszler, "DME from Natural Gas or Biomass: A Better Fuel Alternative," Presentation at SAE Government/Industry Meeting. Washington, D.C., January 2013.

[33] International Energy Agency, 2013. "International EV outlook." Available at http://www.iea.org/publications/freepublications/publication/name,37024,en.html.

[34] Tom Brotherton and Fred Silver, CALSTART, "Cal HEAT Research and Market Transformation Roadmap for Medium and Heavy Duty Trucks: Implications for the GHG/Fuel Economy Standards," Presentation to the committee, July 31, 2013.

sleeper berth. Solutions include battery-operated HVAC and auxiliary power units (APUs), start/stop systems, and truck stop electrification.

Of course, given the range limitations of current vehicle battery technology, electrification is more feasible for some types and modes of MHD vehicles than others. For example, battery-powered motors are least feasible for long-haul heavy-duty trucks that usually travel hundreds of miles per day but may be very promising for service fleets where vehicles perform a number of local deliveries or other jobs per day and then are parked overnight at a centralized base, where they can be plugged in and recharged. One estimate is that up to 6.4 percent of power train systems in MHDVs (including buses) will be electric or hybrid by 2020.[35] This represents slightly over 130,000 units, of which about two-thirds are projected to be hybrids and one-third pure electrics.[36] Other analysts predict that electric and hybrid vehicles will represent only niche markets before 2030, when more significant market penetration is expected.[37]

Another important alternative-fuel technology involves hydrogen fuel cells as the power plant; such fuel cells are projected to significantly penetrate the MHDV sector by the early 2020s. Several light-duty vehicle manufacturers are developing fuel-cell vehicles (FCVs) for commercial introduction, including Hyundai in 2014 and Honda in 2015, with others planning introductions from 2017 to 2020.[38] This will result in technology validation, hydrogen infrastructure development, and cost savings that will eventually benefit the commercialization of FCVs in the MHDV sector. California is supporting the introduction of FCVs through a partnership with vehicle manufacturers and other stakeholders that has developed a roadmap for installing the infrastructure needed for the commercialization of FCVs.[39]

Some MHDV manufacturers currently have active programs developing FCVs. For example, Vision Motor Company has developed the nation's first Class 8 zero emission (tank-to-wheels) hydrogen/electric hybrid vehicle (the Tyrano), designed for local and regional short-haul trips.[40]

Hydrogen fuel cells are also being developed for buses in the MHDV category. For example, the Federal Transit Administration (FTA) has sponsored a cooperative partnership between industry and government to advance the commercialization of fuel-cell technology in U.S. transit buses (FTA, 2012). This program has launched demonstration and evaluation programs for fuel-cell buses in several U.S. cities. For example, AC Transit is currently operating 12 third-generation fuel-cell buses in its HyRoad demonstration program in Oakland, California, that are achieving significantly lower tank-to-wheels fuel consumption than diesel buses while emitting zero pollution.[41] Fuel cells are also being developed to provide auxiliary power for trailer refrigeration, used in some 300,000 refrigerated trucks. The Department of Energy's Pacific Northwest National Laboratory (PNNL) has launched a demonstration project with four trucks whose refrigerated trailers are powered by a fuel cell.[42] According to the PNNL news release, "Industry officials estimate that approximately 300,000 refrigerated trucks with auxiliary power units are on the road in the United States. By replacing the small diesel engines with the more efficient fuel cell, users will see fuel savings of approximately 10 gallons a day per unit, in addition to reduced emissions of pollutants and significantly quieter operation."[43]

The carbon dioxide and fuel consumption benefits of both electric and fuel-cell vehicles will depend to a significant degree on the emissions characteristics of the source used to generate the electricity or hydrogen fuel that powers the vehicle (Babaee et al., 2014).

Life-Cycle Analysis of Fuels

NHTSA and EPA's Phase I Rule, following President Obama's rulemaking request (White House, 2010), considers the fuel consumption of the vehicle and the tailpipe CO_2 emissions that need to be achieved, on average, by the mix of vehicles sold each year by each manufacturer. Manufacturers are likely to achieve these vehicle standards using the variety of different energy fuels and technologies discussed above—including diesel, gasoline, ethanol, natural gas, electric batteries, and fuel cells—each of which will have varying implications for overall GHG emissions and energy demand considered on a life-cycle basis.

While energy consumption and GHG in the end-use phase (onboard fuel consumption and tailpipe emissions) may be the largest contributor in some cases, energy consumption and emissions associated with fuel production, distribution and processing, vehicle efficiency, and end-of-life may contribute to a substantial share of overall vehicle emissions and energy consumption. The committee notes that the 2010 memorandum also states that "[NHSTA and EPA should] propose and take comment on strategies, [...] to achieve substantial annual progress in reducing transportation sector

[35] Sandeep Kar, Frost & Sullivan, "Strategic Outlook of North American Medium and Heavy Commercial Truck Powertrain Market," Presentation to the committee on March 21, 2013.

[36] Ibid.

[37] Eelco den Boer, Sanne Aarnink, Florian Kleiner, and Johannes Pagenkopf, "Zero emissions trucks: An overview of state-of-the-art technologies and their potential, " CE Delft, July 2013.

[38] A. Webb, 2013, "Auto makers renew interest in fuel-cell vehicles: Despite cost, political hurdles." Available at http://wardsauto.com/vehicles-amp-technology/auto-makers-renew-interest-fuel-cell-vehicles-despite-cost-political-hurdles.

[39] http://cafcp.org/carsandbuses/caroadmap.

[40] http://www.visionmotorcorp.com/tyrano.asp.

[41] http://www.actransit.org/environment/the-hyroad./.

[42] PNNL, 2013, "Refrigerated trucks to keep their cool thanks to fuel cell technology," August 23, available at http://www.pnl.gov/news/release.aspx?id=1005.

[43] Ibid.

emissions and fossil fuel consumption consistent with my Administration's overall energy and climate security goals." However, not considering well-to-wheel emissions may lead to regulations that do not achieve the anticipated energy and GHG savings. The committee recognizes there are complexities, uncertainties, and limitations to the feasibility of a regulatory framework such as would expand the scope to such a life-cycle approach from the current end-use approach.

Recommendation 1.10: NHTSA, in coordination with EPA, should begin to consider the well-to-wheel, life-cycle energy consumption and greenhouse emissions associated with different vehicle and energy technologies to ensure that future rulemakings best accomplish their overall goals.

Automated/Connected Vehicles

Significant progress is being made in developing "connected" or "automated" vehicles that can operate more efficiently and safely using advanced telematics technology. To date, most of the progress in this area has focused on light-duty passenger vehicles, with Google and most light-duty vehicle manufacturers actively developing such vehicles for commercialization within the next decade.[44] Although at a slower pace, these technologies will inevitably also be applied to MHDVs. For example, Caterpillar Inc. is currently building 45 automated, 240-ton mining trucks to operate at an Australian iron-ore mine without an onboard operator (Berman, 2013). The most optimistic estimates are that the first automated long-haul trucks (ALHTs) may be commercially viable by the mid- to late-2020s, and could decrease fuel consumption by 15 to 20 percent compared to today's traditional fleets through more efficient driving patterns (Conway, 2013). It is likely that more limited semi-autonomous systems with various driver aids (e.g., adaptive speed control) and enhanced communications (e.g., vehicle-to-vehicle or vehicle-to-infrastructure) will be installed before fully autonomous vehicles are available, and these may provide fuel consumption savings even sooner (Clancy, 2013). Perhaps the most promising technology in the short term is vehicle platooning, in which a set of two or more vehicles is operated closely spaced using semiautonomous intervehicle and navigational communications technologies to reduce aerodynamic drag. Demonstration programs for such vehicle platooning are already under way in the United States,[45] Europe,[46] and Japan.[47] If trucks with semiautomated technologies become available during the compliance period for the Phase II regulations, they may help achieve compliance with the standards, depending on how the standards are structured.

Green Logistics

"Green logistics" refers to innovations in infrastructure, organizational initiatives, or traffic management that can result in more sustainable transport. It may also include increased driver training and other behavioral initiatives. These approaches can result in significant and cost-effective reductions in transport emissions and fuel consumption (Hyard, 2013). Examples of such measures that could impact MHDVs are access control (including lane restrictions), urban traffic control measures, road pricing, smart traffic lights that provide more information to drivers on road conditions and traffic, ramp metering, and other fleet and fuel management approaches. Many U.S. cities and municipalities are actively exploring such programs, which to date have been more widely adopted in Europe and Asia. These, along with operational changes that companies are making to reduce their environmental footprints and improve their bottom lines, may contribute to improved MHDV fuel consumption and reduced CO_2 emissions by the 2020s when the Phase II regulations are in effect.

Background Regulatory Changes

As NHTSA and EPA move forward with the Phase II regulations, it is likely that other federal and/or state regulations will be promulgated that may directly or indirectly affect fuel consumption, GHG emissions, or attempts to control fuel consumption and CO_2 emissions. While the committee has not fully investigated these other regulations, it recognizes that in some cases, they may pose a positive or negative confounding effect on the implementation of Phase 2 MHDV fuel consumption and CO_2 regulations. For example, a more stringent NO_x standard in California may reduce the fuel consumption potential of an MHDV, making compliance with the Phase II Rule more difficult to achieve. A short list of such possible regulations that might interfere with or alternatively assist with compliance with Phase II regulations on fuel consumption and CO_2 emissions is provided in Table 1-1.

[44] KPMG and Center for Automotive Research, 2012. *Self-Driving Cars: The Next Revolution.*

[45] Berkeley University, Partners for Advanced Transportation TecHnology (PATH), available at http://www.path.berkeley.edu/Videos/movie8.html.

[46] Safe Road Trains for the Environment (SARTRE), available at http://www.sartre-project.eu/en/Sidor/default.aspx.

[47] Steven Ashley, "Robot Truck Platoons Roll Forward," BBC Online, April 10, 2013, available at http://www.bbc.com/future/story/20130409-robot-truck-platoons-roll-forward/print/slide/0.

TABLE 1-1 Possible Non-Fuel-Efficiency Regulations in the 2020-2025 Period That Could Affect Fuel Consumption of MHDVs

Possible Regulation	Impact on Fuel Consumption and CO_2 Emissions
California Air Resources Board considering further NO_x emission standards for heavy-duty engines	May increase fuel consumption unless aftertreatment becomes more efficient.
Specifications for new and alternative fuels—e.g., CARB establishing regulations identifying new alternative diesel fuels	Expected not to negatively impact fuel consumption and will improve CO_2 emissions control.
California AB32 (California Global Warming Solutions Act)	Requires statewide plan to reduce GHG emissions by 2020, including provisions that will have direct and indirect benefits in reducing MHDV California emissions relating to fuels, traffic control, and other measures.
Highway speed limits for heavy-duty vehicles	Lower speeds will generally have a significant beneficial impact on fuel consumption, although size of impact will depend on other power demands of the vehicle.
State regulations on fuel exploration, extraction, production, or distribution (including fracking)	Depending on the regulation, natural gas pricing and availability may be negatively affected, which may increase the cost of reductions in fuel consumption.
Wireless roadside inspection programs	Eliminate idling from vehicles having to wait in inspection lines, thereby reducing fuel consumption.
Restrictions on driver distraction and hours	Will promote more efficient trips; beneficial indirect impact in reducing fuel consumption.
Highway funding changes—e.g., VMT fees, fuel tax, weight limits	Many such measures are likely to reduce fuel consumption.
International Harmonized Heavy-Duty Certification Test	Due to the complexity of matters considered in harmonizing regulations (e.g., test cycle, fuel specification, compliance expectations, fuel taxes), the effect of harmonized regulations is uncertain.

Finding: A variety of factors, including alternative fuel technologies, non-fuel-consumption regulatory programs, and other developments will affect the fuel consumption and GHG emissions of medium- and heavy-duty vehicles manufactured and operated in the 2020s.

Recommendation 1.11: In its regulatory analysis, NHTSA should carefully consider and attempt to quantify the impacts of nontechnological factors on the costs and feasibility of future fuel consumption improvements.

REFERENCES

Argonne National Laboratory, Center for Transportation Research. 1999. *Effects of Fuel Ethanol Use on Fuel-Cycle Energy and Greenhouse Gas Emissions.* January.

Babaee, S., A.S. Nagpure, and Joseph F. DeCarolis. 2014. How much do electric drive vehicles matter to future U.S. emissions? *Environmental Science & Technology.*

Berman, D.K. 2013. Daddy, what was a truck driver? *Wall Street Journal*, July 23.

Broder, J., and C. Krauss. 2012. "A big, and risky, energy bet." *New York Times.* December 17.

Button, K. 2010. *Transport Economics*, 3rd ed. Cheltenham, U.K.: Elgar.

Capps, G., O. Franzese, W. Knee, M.B. Lascurain, and P. Otaduy. 2008. *Class-8 Heavy Truck Duty Cycle Project Final Report.* ORNL/TM-2008/122. Oak Ridge, Tenn.: Oak Ridge National Laboratory.

Clancy, H. 2013. How Sensors Promise to Reinvent Truck Driving, Greenbiz.com, August 23.

Conway, P. 2013. The next autonomous car is a truck. *Strategy + Business* 71. May 28.

Cummins, Inc. 2009. Framework for the Regulation of GHGs from Commercial Vehicles.

Energy Information Administration (EIA). 2012. *Annual Energy Outlook 2013: With Projections to 2030.* Washington, D.C.: Energy Information Administration.

Environmental Protection Agency (EPA) (2000). Control of Air Pollution from New Motor Vehicles: Tier 2 Motor Vehicle Emissions Standards and Gasoline Sulfur Control Requirements, Federal Register 65: 6698-6870. February 10.

EPA. 2011. *GHG Emissions Model (GEM) User Guide.* EPA-420-B-11-019. Washington, D.C.: EPA.

EPA and NHTSA. 2011a. Federal Register, Vol. 76, No. 179, pp. 57106 to 57513. September 15.

EPA and NHTSA. 2011b. Final Rulemaking to Establish GHG Emissions Standards and Fuel Efficiency Standards for Medium- and Heavy-Duty Engines and Vehicles: Regulatory Impact Analysis. EPA-420-R-11-901. August. Washington, D.C.: EPA.

Federal Transit Administration (FTA). 2012. FTA Fuel Cell Bus Program: Research Accomplishments through 2011. Available at http://www.fta.dot.gov/documents/FTA_Report_No._0014.pdf. March.

Gaines, L. 2006. Estimation of fuel use by idling commercial trucks, Paper 06-2567. *Transportation Research Record: Journal of the Transportation Research Board.* Washington, D.C.: NRC.

Harrington, W., and V. McConnell. 2003. *Motor Vehicles and the Environment.* Washington. D.C.: Resources for the Future.

Hyard, A. 2013. Non-technological innovations for sustainable transport. *Technical Forecasting and Social Change* 80: 1375-1386.

International Organization for Standardization (ISO). 2009. Passenger car, truck and bus tyres—Methods of measuring rolling resistance—Single point test and correlation of measurement results. ISO 28580:2009. Geneva: International Organization for Standardization.

Knoll, K., B. West, W. Clark, R. Graves, J. Orban, S. Przesmitzki, and T. Theiss. 2009. Effects of Intermediate Ethanol Blends on Legacy Vehicles and Small Non-Road Engines, Report 1–Updated. NREL/TP-540-43543. Golden, Colo.: National Renewable Energy Laboratory.

Lempert, R. 2007. Scenario analysis under deep uncertainty. *Modeling the Oil Transition: A Summary of the Proceedings of the DOE/EPA Workshop on the Economic and Environmental Implications of Global Energy Transitions,* D.L. Greene, ed., ORNL/TM-2007-014.

Merton, R.K. 1936. The unanticipated consequences of purposive social action. *American Sociological Review* 1: 894-904.

National Highway Traffic Safety Administration (NHTSA). 2010. *Factors and Considerations for Establishing a Fuel Efficiency Regulatory Program for Commercial Medium- and Heavy-Duty Vehicles.* DOT HS 811 XXX. Available at http://www.nhtsa.gov/staticfiles/rulemaking/pdf/cafe/NHTSA_Study_Trucks.pdf.

National Renewable Energy Laboratory. 2011. Celluosic ethanol technology on track to being competitive with other transportation fuels. *NREL Highlights*. February.

National Research Council (NRC). 2001. *Effectiveness and Impact of Corporate Average Fuel Consumption (CAFE) Standards.* Washington, D.C.: National Academy Press.

NRC. 2009. *Liquid Transportation Fuels from Coal and Biomass: Technological Status, Costs, and Environmental Impacts.* Washington, D.C.: The National Academies Press.

NRC. 2010. *Technologies and Approaches to Reducing the Fuel Consumption of Medium- and Heavy-Duty Vehicles*. Washington, D.C.: The National Academies Press.

North American Council for Freight Efficiency (NACFE) and Cascade Sierra Solutions (CSS). 2013. *Barriers to the Increased Adoption of Fuel Efficiency Technologies in the North American On-Road Freight Sector*. July.

U.S. Department of Transportation and Federal Highway Administration. 2013. *Evaluation of Ultra-Clean Fischer-Tropsch Diesel Fuel in Transit Bus Applications.* Technical Report FTA-OK-26-7015.2010.1. Washington, D.C.: U.S Department of Transportation. March 31.

White House. 2010. Presidential Memorandum Regarding Fuel Efficiency Standards. Washington, D.C.: Executive Office of the President. May 21.

White House. 2013. *The President's Climate Action Plan*. Washington, D.C.: Executive Office of the President. June.

Yun, J.M. 1997. Measuring the Unintended Effects and Costs of Fuel Consumption Regulation. Emory University.

Zielinski, D. 2013. 2012 tire shipments unchanged. April 3. Washington, D.C.: Rubber Manufacturers Association.

2

Potential for Technological Change in Commercial Vehicles to Impact Future NHTSA Regulations

OVERVIEW OF MEDIUM- AND HEAVY-DUTY VEHICLES (CLASSES 2B THROUGH 8)

Owing to time constraints for preparation of this report, the committee has not been able to conduct a comprehensive analysis of new technologies that would supplement those identified in the National Research Council (NRC) Phase One Report. This will be done in an expanded effort in 2014 to 2015 for the committee's final report, which will include both a thorough update to the data for the Phase One Report technologies and the analysis of any additional new technologies that have emerged since that report's preparation.

The National Highway Traffic Safety Administration (NHTSA) has said that whereas the Phase I Rule was informed by the off-the-shelf technologies in the NRC Phase One Report, the Phase II regulations will be informed by the NRC Phase Two Report on future and advanced technologies. To this end, NHTSA contracted with the Southwest Research Institute (SwRI) to conduct a multiyear study of fuel-efficiency technologies for medium- and heavy-duty vehicles (Classes 2b-8) in the years before and during the Phase II regulations' time frame. The work scope has the following tasks: Task 1: program management; Task 2: literature review and summary tables; Task 3: augment understanding of MY11+ vehicle fleet baseline (engines compliant with 2010 EPA regulations); Task 4: technologies in the MY 2018 fleet; Task 5: technology analysis for Phase II; Task 6: cost-effectiveness analysis for Phase II; Task 7: evaluation of MD/HD fuel economy and emissions testing and simulation approaches; Task 8: technical support—presentations to National Academy of Sciences (NAS)/industry events. Three tasks are of particular interest: Task 4, technologies in the model year (MY) 2018 fleet; Task 5, analysis of potential technologies for Phase II; and Task 6, cost analysis for potential technologies in Phase II. Technologies that SwRI is studying are based on four engines:[1]

- *Detroit Diesel DD15 14.8 L I6.* Optimize turbocompound for cruise performance (sacrifice top-end performance); evaluate electrical turbocompound (decouple power turbine speed from crankshaft speed); remove turbocompound; explore asymmetric turbo concept of 2013 DD15; explore no exhaust gas reduction (EGR) potential, including higher turboefficiency; check lower back pressure and aftercooler pressure differentials; explore reduced parasitic power from water, oil, and fuel pumps; explore engine friction reduction; explore downsizing or downspeeding options, including higher peak cylinder pressure (PCP) if needed; add bottoming cycle; and explore variable valve timing.
- *Cummins ISB 6.7 L I6.* Explore no EGR potential, including higher turbo efficiency; check lower back pressure and aftercooler pressure drop; explore reduced parasitic power from water, oil, and fuel pumps; explore engine friction reduction; explore downsizing or downspeeding options, including higher peak cylinder pressure if needed; possible upsized version for lower Class 8 applications; explore reactivity controlled compression ignition with information provided by the University of Wisconsin; and explore variable valve timing.
- *6.2 liter port injected V-8.* Explore variable valve lift; explore cylinder deactivation (four cylinder); add stoichiometric gasoline direct injection (GDI); add lean burn GDI with selective catalytic reduction (SCR); explore GDI with EGR (high-efficiency dilute gasoline

[1] Thomas Reinhart, Southwest Research Institute, "Phase 2 MD/HD Vehicle Fuel Efficiency Technology Study." Presentation to the Committee on Technologies and Approaches to Reducing the Fuel Economy of Medium- and Heavy-Duty Vehicles, Phase Two. Washington, D.C. June 20, 2013.

engine [HEDGE]); engine downsizing/downspeeding covered by 3.5 v-6; explore friction mean effective pressure (FMEP) improvements.

- *Turbo DI 3.5 V-6.* Explore variable valve lift; explore cylinder deactivation (base engine only); add lean burn GDI with SCR (use experimental data for lean limit and heat release); explore GDI with EGR (HEDGE); engine downspeeding with increased brake mean effective pressure and PCP (4000 rated, 24 bar torque peak); explore FMEP improvements (baseline only); and explore turboefficiency improvement (baseline only).

Although final results from these tasks are still pending, it is anticipated that this study will refine and supplement the NRC Phase One Report findings and recommendations regarding power train and vehicle technologies as summarized in Figure 6-1 and Tables 6-18 and 6-19 of the Phase One Report.

The projections in Figure 6-1 and Tables 6-18 and 6-19 of the NRC Phase One Report were developed under contract by TIAX and reported to the committee in *Assessment of Fuel Economy Technologies for Medium- and Heavy-Duty Vehicles* (TIAX, 2009). In that report TIAX qualified these results with the following statement:

> It needs to be emphasized that our package results are very dependent on the vehicle's duty cycle and as such should be viewed as averages for each truck segment. Further, although the data for this analysis were obtained for the specific truck segments and, in general, for typical duty cycles for these segments, detailed vehicle simulation modeling was not done to establish the fuel consumption benefit of combined fuel savings technologies for each duty cycle. Follow up work is needed to confirm not only the individual estimated benefits but also the packaged benefits.

Neither the NRC Phase One Report nor NHTSA's *Factors and Considerations for Establishing a Fuel Efficiency Regulatory Program for Commercial Medium- and Heavy-Duty Vehicles* (2010) cited this important qualification.

Finding: There is a wide range of possible fuel consumption potentials for various technologies and, further, a high likelihood of interactions between the latter when applied in combination such as might impact the aggregate fuel consumption potential. There is a wide range of ways in which various technologies could reduce consumption. There is also a high likelihood that the technologies will interact when they are applied in combination to further reduce aggregate fuel consumption.

Recommendation 2.1: NHTSA should conduct detailed simulation modeling and physical testing of various logical technology combinations and use the results to guide the setting of Phase II regulations.

In establishing the stringency of Phase II regulations, careful evaluation of technology penetrations is necessary. While NHTSA said that the intent of the Phase I Rule was to base the technologies on off-the-shelf technologies, the new baseline for revising those regulations is technologies embodied in more current vehicles, which will include not only off-the-shelf technologies but also those future/advanced technologies that will have penetrated the market by 2018. SwRI project Task 3, MY2011+ fleet baseline, should show new technology penetrations that could become an updated baseline from which the Phase II Rule can be projected, depending on its success in capturing advances in the 2014-2018 time frame.

Recommendation 2.2: NHTSA's Phase II Rule should take the current and projected incremental fuel consumption reductions and penetration rates of the various technologies into careful consideration: These incremental reductions and penetration rates should be updated from those that were projected in the Phase I rulemaking. Furthermore, system interactions should be evaluated for the effect on the projected incremental reductions whenever combinations of technologies are considered.

NRC (2010), NHTSA (2010), and TIAX (2009) (the latter prepared for the NRC Phase One committee) all included extensive descriptions, evaluations, and projections for the various individual power train and vehicle technologies identified in the Phase One Report. The results were summarized in Tables II.C.1 through II.C.9 of NHTSA (2010) and were derived from TIAX (2009). Given that these estimates are now more than 4 years old, high priority should be given to determining current values for these technologies and for including any new technologies that have emerged in the interim. The SwRI study for NHTSA should provide meaningful updates to these estimates. Additionally, the committee has to date met with and heard presentations from 32 government agencies, nongovernment organizations, and companies that sell automotive subsystems, engines, transmissions, and vehicles in Class 2b through Class 8 (see Appendix C). These categories were represented by presentations from Daimler, Volvo, Navistar, Cummins, Westport, Eaton, General Motors (GM), Ford, and others. To augment these presentations to the full committee, small groups of committee members visited original equipment manufacturers (OEMs) at their facilities or had telephone conversations with them. These activities are ongoing as the need for information arises.

The costs of these various technologies are germane as well. In the process of assessing the cost of new technologies, the Regulatory Impact Analysis (EPA and NHTSA, 2011) presented cost estimates for selected technologies for MY2030 and MY2050 compared to the 2011 baseline cost (in additional U.S. dollars). The committee has asked OEMs to comment on the order of magnitude of these increases;

however, given the fast track for this first report, these comments are pending. Likewise, projections of cost per truck and annual costs based on truck sales to 2050 are still under deliberation.

The following paragraphs will reflect only new technologies, fuel consumption estimates, or other considerations that have been revealed to the committee to date as potentially available in the Phase II time frame. As in NRC (2010; especially on p. 4, Tables S-1 and S-2), the committee addresses the uncertainty inherent in these estimates by reporting a range of values for the percentage reduction in fuel consumption that might be attributed to the addition of a single technology. The committee also analyzes the sensitivity of these estimates to certain variables such as environmental temperature (e.g., cold-start conditions), vehicle duty cycle, and/or manufacturing tolerances, and notes the effect where it is significant. Further, as noted above and quoted from TIAX (2009), the simple method of aggregating the single technologies may provide a different result than what a specific combination of technologies would produce in a power train or whole vehicle as a system. This system-induced uncertainty must also be recognized and evaluated through either simulation or physical testing of technology combinations.

POWER TRAIN TECHNOLOGIES

Among the new engine and vehicle technologies that have been introduced since the Phase One Report, the emergence of natural gas as a transportation fuel is of significance. In addition, the growing interest in dimethyl ether (DME) as a finished fuel and the availability of natural gas and related renewable and alternative feedstocks to produce DME justify a focused analysis of technologies for natural gas engines and for other related vehicles. The latter topic will be covered further in Chapter 6.

All the OEMs are embarked on developing the engine. Among the specifics they are addressing are the following:

- *Turbocharging, including dual-stage turbocharging with intercooling, mechanical and electric turbocompounding, and advanced EGR cooling.* No additional new technologies or estimates have been provided in the reviews to date.
- *Electrification of engine accessories.* Traditional belt- or gear-driven accessories can be converted to electric power. The fuel consumption reduction results from the fact that in an electric actuation some accessories (such as power steering and the air compressor) can be operated only when needed. Other accessories (such as the water pump or cooling fan) could be run at speeds independent of engine speed. Either of the two cases can reduce fuel consumption. Electrification of accessories provides a 3 to 5 percent fuel consumption reduction if applied as a package on a hybrid vehicle. This benefit is more effective in urban driving conditions and in short-haul use; line-haul applications will benefit less. Vehicle and engine manufacturers of Class 2 through Class 7 vehicles report that electrical power steering is already in place and undergoing a global migration.
- *Reduction of engine friction.* Engine friction reduction has been pursued continuously by manufacturers through careful design and selection of advanced materials. Further efforts to make reductions face the added challenge of avoiding issues with durability and poorer performance. In the recent past, attention was given to using thinner lubricants; lower viscosity oil such as 10W30 was tested as a replacement for the standard 15W40 oil. It has now been confirmed by testing that a further reduction of 1 to 1.5 percent in fuel consumption may be obtained with thinner oils once durability has been confirmed. Thermostatic control of oil cooler—a solution used selectively in the past—can maximize lubricant performance over a broad temperature range. Some testing has reported a reduction in fuel consumption closer to 2 percent. The effect is more pronounced for cold starts and low-load operation. The introduction of greater volumes of synthetic base stocks will allow the evaluation and possible use of even lower viscosity oil formulations such as 5W30 and 0W30. In combination with advanced friction modifiers and viscosity improver additives, this could provide additional fuel efficiency. The amount of improvement, durability considerations, and penetration expectations will be evaluated and considered for the committee's second report, due in 2016.
- *Improvement of diesel exhaust particulate matter (PM) control using a diesel particulate filter (DPF) with a catalyst coating.* The use of a DPF can degrade the fuel efficiency of the engine owing to exhaust flow restriction and pressure buildup in the system or the need for additional fuel to maintain the operation of the DPF. PM is collected in the DPF and disposed of periodically, either by combustion when the temperature is high during the normal cycle of operation or by injecting diesel fuel into the exhaust upstream of the DPF at light-load engine operation. A diesel oxidation catalyst oxidizes the fuel and generates the heat that regenerates the DPF in some manufacturers' emission solutions. The current state of technology for DPFs is such that the additional fuel used ranges from 0 to 4 percent[2] due to the fuel used for regeneration and the increased back pressure over the duty cycle.
- *Improvement of diesel exhaust catalytic system efficiencies using selective catalytic reduction (SCR).* Control of nitrogen oxide (NO_x) has been accomplished with

[2] W. Addy Majewski, Filters Regenerated by Fuel Combustion, Dieselnet.com.

cooled EGR and an SCR catalyst. The SCR catalyst uses the injection of a urea solution, diesel emission fluid (DEF), for converting NO_x into nonharmful compounds. The infrastructure buildup for SCR/DEF was a significant challenge before 2010; however, the technology solution was ready for the 2010 federal NO_x standards. This use of DEF is calculated as the equivalent fuel for consideration of the total fuel used. The 2010 emission standards for NO_x required the use of SCR, but most manufacturers reduced the amount of EGR used and thereby improved the fuel efficiency of the engine (6.5 percent).[3] The net reduction in fuel consumption (engine efficiency minus DEF usage) is 2 to 4 percent.[4,5] This was one of the few heavy-duty emission control technologies in history that simultaneously enabled increased efficiency and reduced emissions. Encouraging new catalyst technology directions employing nanoscale catalytic materials on inert substrates are under investigation and will be evaluated by the committee in its second report.

At present, the above systems for NO_x and PM control are working satisfactorily. Small improvements continue to be achieved, but the technology has attained maturity, and the fuel efficiency penalty may stabilize at about 3-4 percent for the entire system or about 2 percent for the PM control. These figures should be checked again as part of discussions with OEMs for the final report of the committee.

- *Improvement in thermal efficiency of the diesel cycle in real operating conditions and with real engines (rather than in a controlled laboratory prototype).* The challenge of improving thermal efficiency of the diesel cycle has been an active topic for research for several years. In a laboratory environment and under very controlled conditions, researchers can reach a thermal efficiency of 50 percent. They achieved this by simultaneous improvement of the compression ratio, expansion ratio, combustion chamber architecture, injection timing, injection pressure, rate shaping, air: fuel ratio control, air/fuel mixing, etc. All these measures contribute to a higher combustion temperature, which is good for lower fuel consumption, but they also increase NO_x emissions, which is not good. It is expected that fuel injection may evolve further so that injection pressures that were as high as 2,300 bar in 2010 may increase to 3,000 bar by 2015 and perhaps to 4,000 bar by 2020. Moreover, new injection techniques may emerge (such as supercritical injection) and piezoelectric nozzle use may increase. A reduction of up to 6 percent in fuel consumption may be realizable. Also, if real-time combustion control becomes available with start of combustion sensors, then an additional 1 to 4 percent reduction in fuel consumption can be expected.
- *The use of alternative combustion processes such as homogeneous charge compression ignition (HCCI), premixed charge compression ignition (PCCI), and low-temperature combustion (LTC), even if in some limited operational range, for emission reduction.* In search of improved cycle efficiency and, especially, generation of very low NO_x levels, several combustion processes have been researched, such as HCCI, PCCI, and LTC. Some success has been obtained, but many challenges remain, especially in control over the entire range of engine operation. In the interim, approaches that provide partial operation of the engine in alternative modes are under investigation or have been demonstrated. Reactivity-controlled compression ignition (RCCI) has been shown to offer better controllability of premixed-type combustion using two fuels of different ignition properties. It is under investigation by a number of organizations.
- *Improved efficiency of the driveline and improved system integration using strategies that enable the engine to operate at higher drive-cycle efficiency.* Several technologies are worthy of mention. One is waste heat recovery. Cummins' Supertruck demonstration with integrated bottoming cycle has achieved ~50 percent on-road efficiency. For Class 2b vehicles, the potential for 10-speed automatic transmissions with approximately 1 percent fuel consumption improvement compared to current 6-speed automatics has been identified by one OEM, GM. Cost and availability have not been provided, and it has been classified as High Development Risk. Another transmission manufacturer, Eaton, has shown a dual-clutch automatic transmission for line haul and vocational applications that reduces or eliminates power interruption during shifts. Allison has introduced a new 10-speed automatic transmission with a measured improvement in fuel economy compared to manual and AMT transmissions in Class 8 day cab vehicles.
- *Hybrid power trains, including regenerative braking, engine downsizing, engine shut-off, enabling electrification accessories, plug-in hybrids, etc.* No additional new hybrid systems have been identified in the reviews to date. However, given the high duty-cycle dependency, energy storage methods, costs, and relatively large potential fuel consumption reductions projected across most vehicle classes, NHTSA should form a study focused in this area to identify current realistic penetration rates and appropriate simulation and test methodologies to determine the resulting potential for fuel consumption reduction. Several manufacturers pointed out that with the ever more rapid rates at which new energy sources and new energy storage technolo-

[3] W. Addy Majewski, SCR Systems for Mobil Engines, Dieselnet.com.

[4] Ibid.

[5] NRC (2010).

gies are being adopted, the points of regulation and the certification methodologies need to be examined and potentially modified to more accurately evaluate and credit this trend. Improvements to be evaluated included propulsion system dynamometer certification instead of engine-only certification; more emphasis on transients in modeling, simulation, and testing; and standards and certification only at the vehicle level.

VEHICLE TECHNOLOGY

The following technologies involving the other components of the vehicle are addressed by the OEMs as identified in the Phase One Report:

- *Aerodynamic losses at high vehicle speed; improvements critical for vehicle aerodynamic optimization.* No additional new technologies have been identified to date. Two truck OEMs (Daimler[6] and Navistar[7]) have indicated that alternatives to coast-down testing should be considered, such as full-scale and scaled-down wind tunnel testing and computational fluid dynamics (CFD) analysis. They also indicated that aerodynamic bins should be based on wind-averaged drag rather than zero-degree-yaw drag. Daimler also suggested that greater accuracy would result by narrowing the Phase I Rule's aerodynamic bins and increasing the number of bins by three.
- *Implementation of aerodynamic features; barriers to implementation; cost and robustness at low speeds.* No additional new technologies have been identified to date, but continued development has yielded more robust aerodynamic trailer skirts.
- *Rolling resistance.* The rolling resistance accounts for about 30 percent of the power to move a line-haul truck on level roads and at highway speeds. Reductions in the coefficient of rolling resistance of tires have been enabled both by development of new tire designs for standard width tires and the introduction of wide-base single tires (WBSTs). This technology allows the replacement of two conventional dual tires with a single new-generation wide tire; the coefficient of rolling resistance can be lowered from between 5 and 8 kg/ton to 4 or 5 kg/ton. Real-world testing and modeling have estimated almost a 10 percent improvement in fuel economy from this technology. In addition, some earlier studies by the Environmental Protection Agency (EPA) have also demonstrated reduction in oxides of nitrogen (Bachman et al., 2005). It is to be noted that the tire resistance is also influenced by tire wear (tread depth), by drive cycle, and by load.

Optimizing tire performance for a particular use is challenging because the numerous requirements are sometimes met by contradictory tire characteristics. It might therefore be impossible to have low-rolling-resistance tires for all vehicular applications. Fleets generally recycle partially worn tires by moving them from drive axles to the trailer, resulting in a less than optimum total vehicle package.

WBSTs provide a weight saving compared with dual tires of about 340 kg per five-axle combination tractor-trailer rig. This allows an increase in payload capacity, and it can improve freight efficiency. Despite these advantages, the adoption of WBST is limited owing to concerns about the occurrence of flats; stability and safety of the vehicle in the event of tire failure; availability of replacement tires; and the damage to roads caused by tire failure. Future discussions and test experience may show whether the effect of tire failure on safety has been overstated due to insufficient real data. At least one analysis (TIAX, 2009) believes that by 2016, the new tire technology applied to more axles may bring about a reduction in fuel consumption of about 11 percent in long-haul trucks.

- *Vehicle mass; vehicle lightweighting.* The truck weight impacts the power needed to move the vehicle through rolling resistance, climbing grades, and accelerations. Use of lightweight materials and structures, such as cab structures, wheels, fifth wheel, bell-housing, etc., have contributed to reducing weight in tractors; additionally, aluminum composite panels have reduced the weight of trailers. A barrier to further reduction is the higher cost of light materials. Lightweighting is simultaneously balanced by the increase in vehicle mass needed to accommodate additional systems and equipment, such as new emission control equipment, aerodynamic improvement equipment, waste heat recovery, and hybrid components. No additional new technologies have been identified to date.

[6] Mike Christianson, Daimler Trucks North America, "OEM experience with GHG Phase I and recommendations for Phase II," Presentation to the committee, June 20, 2013.

[7] Greg Fadler, Navistar, Inc., "Navistar fuel economy and emissions," Presentation to the committee, March 20, 2013.

REFERENCES

Bachman, L., A. Erb, and C. Bynum. 2005. Effect of single wide tires and trailer aerodynamics on fuel consumption and NO_x emissions of Class 8 line-haul tractor-trailers. SAE Paper 2005-01-3551. Warrendale, Pa.: SAE International.

EPA and NHTSA. 2011. *Final Rulemaking to Establish GHG Emissions Standards and Fuel Efficiency Standards for Medium- and Heavy-Duty Engines and Vehicles: Regulatory Impact Analysis*. EPA-420-R-11-901. Washington, D.C.: August.

National Highway Traffic Safety Administration (NHTSA). 2010. "Factors and considerations for establishing a fuel efficiency regulatory program for commercial medium- and heavy-duty vehicles." DOT HS 811 XXX. http://www.nhtsa.gov/staticfiles/rulemaking/pdf/cafe/NHTSA_Study_Trucks.pdf.

National Research Council (NRC). 2010. *Technologies and Approaches to Reducing the Fuel Consumption of Medium- and Heavy-Duty Vehicles.* Washington, D.C.: The National Academies Press.

TIAX. 2009. *Assessment of Fuel Economy Technologies for Medium- and Heavy-Duty Vehicles*. Report prepared for the National Academy of Sciences by TIAX LLC. Cupertino, Calif. July 31.

3

Certification and Compliance Procedures Using GEM

This chapter will provide information on the development of the Greenhouse Gas Emissions Model (GEM) for vehicle certification of Class 4 through Class 8 vehicles, a description of the model, use of the model by vehicle original equipment manufacturers (OEMs), and analysis of the model with discussion of some of its limitations. Certification of Class 2b and Class 3 vehicles using a full chassis dynamometer is not addressed in this report, nor are changes to engine certification per Environmental Protection Agency (EPA) requirements for criteria pollutants, greenhouse gases (GHGs), and fuel use. Both subjects will be addressed in the final report.

From both technical and operational perspectives there is an objective to improve, and even optimize, efficiency metrics for trucks in each class. The existing National Highway Traffic Safety Administration (NHTSA) and EPA rules have already addressed the units to be employed for measuring engine efficiency, vehicle efficiency, and the associated carbon dioxide emissions levels, for both gasoline and diesel. From a practical standpoint, at the time of writing, the rules also drive change that provides economic benefit to the truck user. However, as truck efficiency standards advance, there are trade-off issues that must be addressed. The first is central to this report.

- *Fuel efficiency versus greenhouse gases.* If the change or optimization addresses reduction in fuel use, it may not necessarily address reduction in GHGs when a spectrum of fuels is considered. Different fuels have different carbon content, and they may also be associated with production of other GHG species. If fuel-insensitive efficiency and GHG rules are promulgated separately, two rules could drive toward two different fuel or technology options. For example, present-day natural gas engines may benefit GHG reduction more than diesel engines, but the inverse is true when it comes to efficiency. Dual-fuel engines increase the complexity because the balance of fuel use may be activity dependent.

There are also trade-offs with other national interests:

- *Fuel efficiency or greenhouse gases versus cost.* Reducing fuel use or GHG emissions may not be the most economically attractive scenario. Technology costs in some cases may exceed fuel savings over the vehicle life, and the least expensive fuel and technology combination may not offer the best efficiency or lowest GHG scenario. At a higher level, fuel choices may have substantial economic impacts beyond the trucking industry. For example, an advanced aerodynamic device that offers drag reduction of less than 1 percent is unlikely to offer payback during the first period of ownership if the weight and cost cross a certain threshold.
- *Fuel efficiency or greenhouse gases versus criteria pollutants.* Although reducing engine power output tends to reduce both fuel mass rate and criteria pollutant mass rate, some technologies designed to reduce criteria pollutants have adverse effects on engine efficiency and GHG production. In the technical realm, there is often a trade-off between these metrics. For example, a diesel particulate filter will reduce particulate mass but will raise fuel consumption due to exhaust back pressure and fuel used for necessary regeneration.
- *Energy security versus efficiency and emissions.* The use of alternative energy resources or a balancing of source uses may not yield highest efficiency, lowest GHG, or lowest criteria pollutants, but it may satisfy compelling national needs. For example, natural gas, as a domestic fuel, displaces imported oil. However, a spark-ignited natural gas engine is generally less energy efficient than a diesel engine.
- *Technology impact.* Scenarios may be made more complex when one fuel can be used in engines or power trains that employ two fundamentally different technologies if one technology offers benefit for one

metric and the other offers benefit for another metric. For example, natural gas may be used in compression-ignited or spark-ignited modes.

These metrics all have different currencies, and it is impossible to establish exchange rates between them from purely technical arguments. The balancing of these metrics is an issue of policy.

Recommendation 3.1: NHTSA, in consultation with EPA, should consider carefully the impact on related metrics when attempting to optimize for a single metric, or should otherwise establish a clearly articulated objective that weights, or places limits upon, relevant metrics.

DEVELOPMENT OF GREENHOUSE GAS EMISSIONS MODEL

The predecessor to the present committee, the National Research Council (NRC) Committee to Assess Fuel Economy Technologies for Medium- and Heavy-Duty Vehicles, published its final report (the "Phase One Report") in 2010 (NRC, 2010). Appendix G of that report summarizes information on vehicle simulation tool requirements for regulatory use. Key requirements were outlined (NRC, 2010, pp. 221-225); they included

- Maximum reusability,
- Maximum flexibility,
- Selectable complexity,
- Code neutrality,
- Graphical user interface
  —Select architecture, model, and data,
  —Check model compatibilities to avoid crash or erroneous results,
- Select simulation type, including component evaluation, vehicle fuel efficiency, or drive quality
  —Results visualization,
- Generic processes,
- Linkage with other tools,
- Database
  —User access control,
  —Enterprise-wide solutions,
  —Version control,
  —Database search, and
- Selection of single versus multiple tools for regulation.

The final rule was based, in part, on Recommendation 8-4 of the NRC Phase One Report.

> **Recommendation 8-4.** Simulation modeling should be used with component test data and additional tested inputs from power train tests, which could lower the cost and administrative burden yet achieve the needed accuracy of results. This is similar to the approach taken in Japan, but with the important clarification that the program would represent all of the parameters of the vehicle (power train, aerodynamics, and tires) and relate fuel consumption to the vehicle task. Further, the combined vehicle simulation/component testing approach should be supplemented with tests of complete vehicles for audit purposes. (NRC, 2010, p. 190)

GEM Version 1, the proposed simulation tool, was provided to the industry for peer review. Comments were received and discussed in "Peer Review of the Greenhouse Gas Emissions Model (GEM) and EPA's Response to Comments" (EPA, 2011a), which was published in August 2011, just prior to publication of the Phase I Rule. In addition, EPA (2011b) published a document to accompany the rulemaking, *Greenhouse Gas Emissions Standards and Fuel Efficiency Standards for Medium- and Heavy-Duty Engines and Vehicles: EPA Response to Comments Document for Joint Rulemaking*. This document included comments made by the OEMs and others on GEM and EPA's responses. Some corrections to GEM were completed, and GEM Version 2.0 was released.

On September 15, 2011, EPA and NHTSA published their Final Rule on Greenhouse Gas Emissions Standards and Fuel Efficiency Standards for Medium- and Heavy-Duty Engines and Vehicles. GEM is a key part of the final rule. According to the Federal Register of that date,

> compliance with the vehicle standard will typically be determined based on a customized vehicle simulation model, called the Greenhouse gas Emissions Model (GEM), which is consistent with the NAS Report recommendations to require compliance testing for combination tractors using vehicle simulation rather than chassis dynamometer testing. It is an accurate and cost-effective alternative to measuring emissions [of carbon dioxide] and fuel consumption while operating the vehicle on a chassis dynamometer as an indirect way to evaluate real-world operation and performance, various characteristics of the vehicle are measured and these measurements are used as inputs to the model. These characteristics relate to key technologies appropriate for this subcategory of truck—including aerodynamic features, weight reductions, tire rolling resistance, the presence of idle-reducing technology, and vehicle speed limiters. The model also assumes the use of a representative typical engine, rather than a vehicle-specific engine, because engines are regulated separately. Using these inputs, the model will be used to quantify the overall performance of the vehicle in terms of CO_2 emissions and fuel consumption. The model's development and design, as well as the sources for inputs, are discussed in detail in Section II below and in Chapter 4 of the RIA [Regulatory Impact Analysis]. (57 Federal Register 57116)

GEM is employed for model year (MY) 2014 vehicles and later. Given the rules for determining the model year of a vehicle, a 2014 model vehicle could be produced as early as January 2, 2013, or as late as December 31, 2014 (see Box 3-1).

BOX 3-1
EPA's Regulations for Determining Model Year Designation

§ 85.2305 Subpart X—Determination of Model Year for Motor Vehicles and Engines Used in Motor Vehicles Under Section 177 and Part A of Title II of the Clean Air Act.

§ 85.2301 Applicability. The definitions provided by this subpart are effective February 23, 1995 and apply to all light-duty motor vehicles and trucks, heavy-duty motor vehicles and heavy-duty engines used in motor vehicles, and on-highway motorcycles as such vehicles and engines are regulated under section 177 and Title II part A of the Clean Air Act.

§ 85.2302 Definition of model year. Model year means the manufacturer's annual production period (as determined under § 85.2304) which includes January 1 of such calendar year, provided, that if the manufacturer has no annual production period, the term "model year" shall mean the calendar year.

§ 85.2303 Duration of model year. A specific model year must always include January 1 of the calendar year for which it is designated and may not include a January 1 of any other calendar year. Thus, the maximum duration of a model year is one calendar year plus 364 days.

§ 85.2304 Definition of production period. (a) The "annual production period" for all models within an engine family of light-duty motor vehicles, heavy-duty motor vehicles and engines, and on-highway motorcycles begins either: when any vehicle or engine within the engine family is first produced; or on January 2 of the calendar year preceding the year for which the model year is designated, whichever date is later. The annual production period ends either: When the last such vehicle or engine is produced; or on December 31 of the calendar year for which the model year is named, whichever date is sooner. (b) The date when a vehicle or engine is first produced is the "Job 1 date," which is defined as that calendar date on which a manufacturer completes all manufacturing and assembling processes necessary to produce the first saleable unit of the designated model which is in all material respects the same as the vehicle or engine described in the manufacturer's application for certification. The "Job 1 date" may be a date earlier in time than the date on which the certificate of conformity is issued.

SOURCE: 49 CFR 565.15 and 60 Fed. Reg. 4738; January 24, 1995, unless otherwise noted.

Validation of the GEM was reported on the Regulatory Impact Analysis (EPA and NHTSA, 2011) as follows:

- Verification against two chassis dynamometer full combination-vehicle tests at Southwest Research Institute (SwRI) with 2 percent accuracy against three test cycles.
- Verification to another vehicle simulation tool, GT-Drive,[1] with 2 percent accuracy. Ten vehicles were run against three test cycles.

On June 17, 2013, NHTSA and EPA published technical amendments to the rule in the Federal Register (78 Fed. Reg. 36370). Minor changes to the procedures for rounding in the GEM were provided, especially to align the different requirements of EPA and NHTSA. This is intended to make certain that a vehicle ends up in one, and only one, vehicle family/subfamily in both EPA and NHTSA designations. Coincident with this, NHTSA and EPA released GEM version 2.0.1 to the industry. Additionally, changes were made to the multiplication factor for advanced technology credits. This did not affect the GEM calculations or procedures for the baseline vehicle. Another change relates to the use of automatic engine shutdown (AES) technologies. EPA assumes 1,800 hours per year of idling. The technical amendment provides a requirement to discount the effect of the AES if it does not prevent 1,800 hours per year of idling. An error in coast-down testing done in preparation for the final rule was introduced into the definition of the aerodynamic bins.[2] This was corrected in the technical amendment. Since this is an input to the GEM, it has a minor effect on the use of GEM in determining final results.

On August 16, 2013, NHTSA and EPA withdrew several of the technical amendments published on June 17, 2013.

[1] The Phase I Rule references "GT-Power," perhaps spuriously (76 Fed. Reg. 57146).

[2] Binning is a method employed in model development to represent the real-world characteristics of a vehicle in a stylized, discretized manner. It involves the creation of a predefined set of notional categories into which real-world vehicles are sorted for purposes of carrying out the simulation. For example, GEM utilizes five bins to represent the aerodynamic characteristics of various vehicle configurations (EPA and NHTSA, 2011b, p. 2-46).

However, none of the withdrawals pertained to the use of GEM or the new version of GEM with revised rounding.[3]

The cooperative agreement between NHTSA and the NRC, modified in September 2013, includes the following element of the amended statement of task (Appendix B) that is of relevance to this chapter:

> The committee will analyze and provide options for improvements to the certification and compliance procedures for medium- and heavy-duty vehicles—including the use of representative test cycles and simulation using various models—such as might be implemented in revised fuel consumption regulations affecting MY 2019-2022.

The committee has talked with several but not all vehicle manufacturers that use GEM. Based on those conversations, on its review of GEM, and on the need to acknowledge efficient technology broadly in a simulation model, the committee makes several recommendations in the present report. Execution of GEM requires insertion of data on aerodynamic properties and tire rolling resistance. These data are obtained from measurements, which are also discussed below.

BOX 3-2
Criteria Pollutants

The standards for criteria pollutants are found at Title 40 Code of Federal Chapter 1037.102 entitled "Exhaust emission standards for NO_x, HC, PM, and CO." These pollutants are sometimes described collectively as "criteria pollutants" because they are either criteria pollutants under the Clean Air Act or precursors to the criteria pollutant ozone. These pollutants are also sometimes described collectively as "non-greenhouse gas pollutants," although they do not necessarily have negligible global warming potential. As described in § 1037.102, standards for these pollutants are provided in 40 CFR part 86.

DESCRIPTION OF GEM

The GEM model was developed for NHTSA's and EPA's Phase I Rule on medium- and heavy-duty vehicles (MHDVs) as a simplified method for determining the effects of the vehicle (rather than the engine) on fuel economy and GHG emissions. There are separate regulations focused on a certification of the engine as meeting established criteria for carbon dioxide (CO_2), methane (CH_4), and nitrous oxide (N_2O). As such, it builds upon the work already done by EPA and the established procedures in 40 CFR Part 1065 to certify engines in a test cell for the criteria pollutants carbon monoxide (CO), nitrogen oxides (NO_x), hydrocarbons, and particulate matter (see Box 3- 2). The executable GEM program is based on MATLAB/Simulink,[4] a common language for modeling and simulation in engineering. The user has five main inputs to the model:

- Aerodynamic drag coefficient value by bin,
- Steer and drive tire rolling resistance by value,
- Vehicle speed limiter presence and value,
- Vehicle mass reduction (lb), and
- Idle reduction presence.

Additional opportunities are classified as Innovative Technologies, which require a separate procedure and submittal process for taking advantage of the credits allowed. Hybrid technology is included in the Advanced Technology and Innovative Technologies categories. Natural-gas-powered vehicles, which will replace a fraction of the diesel-powered vehicles and which will be important in the coming years, are also included in the Advanced Technology and Innovative Technologies categories. Chapter 4 of the Regulatory Impact Analysis (RIA) that accompanied the promulgation of the Phase I Rule (EPA and NHTSA, 2011) provides an extended explanation of the model.

A brief overview is obtained by looking at the input screen of the model, shown in Figure 3-1. The output of GEM is cycle-weighted g/ton-mi CO_2 and gal/1,000 ton-mi. An important part of understanding the model consists in examining the underlying assumptions, including the following:

- Fixed engine fuel map for one fuel type, diesel;
- Fixed manual transmission (10-speed for Classes 7 and 8 and 6-speed for vocational);
- Fixed axle ratios;
- Fixed tire sizes and a rolling resistance that is invariant with respect to speed and torque, though user supplied;
- Fixed electrical load;
- Fixed mechanical accessory power;
- Three fixed cycles (California Air Resources Board [ARB] Transient, 55 mph Cruise, 65 mph Cruise);
- Level roads;
- 1,800 hours engine idle;
- A fixed payload, although augmented with a weight reduction option; and
- A value for the product of area and drag coefficient, which is a user-supplied constant.

Since GEM is a relatively simple model focused on aerodynamics, rolling resistance, speed, weight, and idle control, it is not capable of acknowledging efficiency or

[3] This change was "to address rounding inconsistencies when converting CO2 values to equivalent fuel consumption values in the Greenhouse Gas Emissions Model (GEM) simulation tool" (76 Fed. Reg. 36377).

[4] For further information, see http://www.mathworks.com. Accessed November 14, 2013.

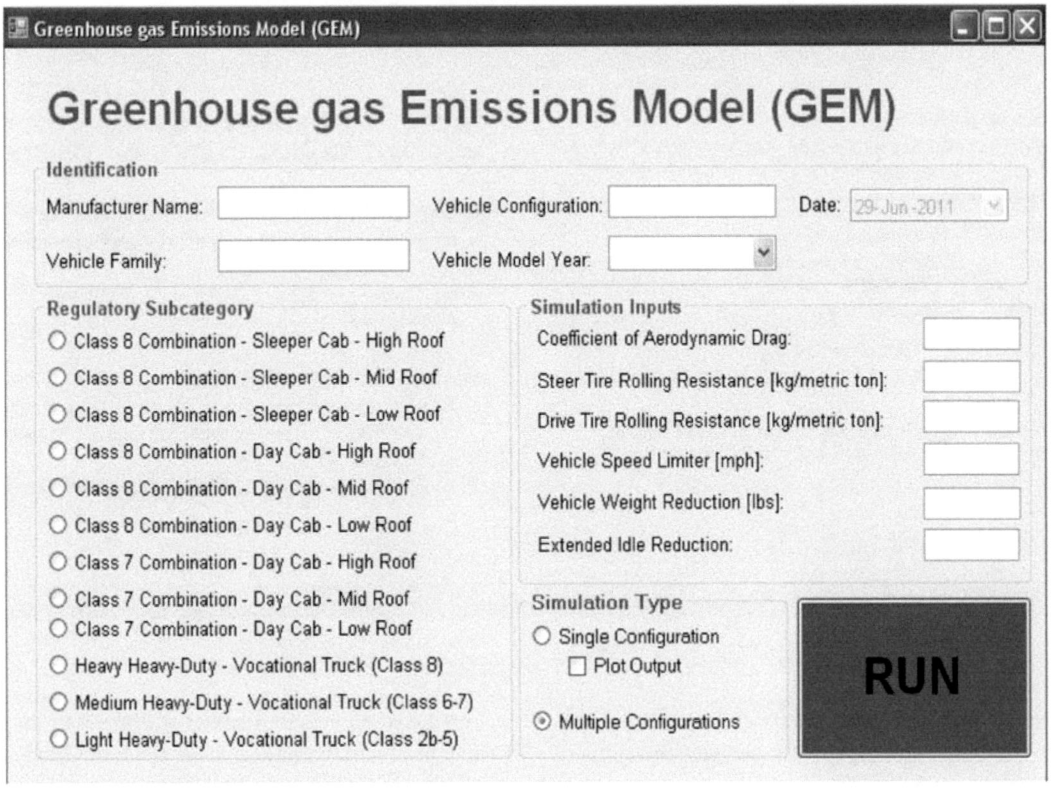

FIGURE 3-1 Graphical user interface for GEM. SOURCE: EPA.

GHG emissions changes associated with the integration of advanced power trains, alternative fuels, hybrid and electric vehicles, and optimal component management. While tire rolling resistance is a key input, the assumption in GEM is that the tires are properly inflated per the recommendation of the tire manufacturer. In the real world, tires are not always maintained at the proper pressure. Tire pressure monitoring, maintenance, and control can only be addressed as Incentivizing Technology. Idle reduction input to the model is limited to a yes/no type answer, whereas idle reduction strategies in actual use can be much more sophisticated, especially as the latter relates to safety of the driver in hot and cold weather.

Certain items other than the few inputs to the model are handled as Incentivizing Technologies. NHTSA describes these in a presentation, shown in Figure 3-2.

To the committee's knowledge, there had been little use of the track for Advanced Technology credits as of October 2013. Discussions with OEMs—Ford, GM, Navistar, Daimler Trucks, PACCAR, and Cummins—suggest three reasons for this:

1. Volumes of product with the potential for Incentivizing Technology are low at this time;
2. 2014-2016 requirements can likely be met without the need for Incentivizing Technology, while the need in 2017 is still open; and
3. The procedure for proving effectiveness of the Incentivizing Technology is burdensome.

USE OF GEM FOR OVER-THE-ROAD TRACTORS

For reporting purposes, manufacturers are obliged to execute GEM to cover sales of Class 7 and 8 on-road tractors. The process is similar for vocational vehicles, but aerodynamic drag is not considered for those vehicles. Exception is granted for a limited number of vocational tractors, which are treated as vocational vehicles, as detailed in the following:

> Class 7 and Class 8 tractors certified or exempted as vocational tractors are limited in production to no more than 21,000 vehicles in any three consecutive model years. (78 Fed. Reg. 36403)
>
> Vocational tractors generating credits can trade and transfer credits in the same averaging sets as tractors and vocational vehicles in the same weight class. (78 Fed. Reg. 36403)
>
> Off-road operation. Heavy-duty vocational vehicles including vocational tractors meeting the off-road criteria in 49 CFR 523.2 are exempted from the requirements in this paragraph (b), but the engines in these vehicles must meet the requirements of paragraph (d) of this section. (76 Fed. Reg. 57499)

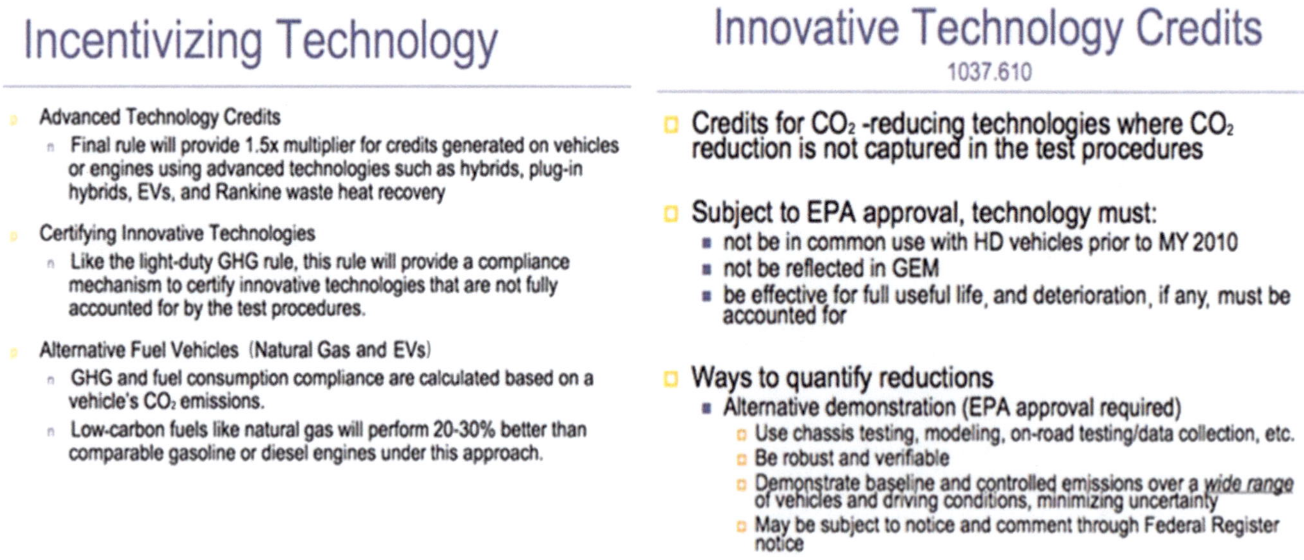

FIGURE 3-2 Inclusion of incentivizing technologies in the Phase I Rule. SOURCE: NHTSA.

GEM is used by the manufacturer to compute projected load-specific fuel consumption for nine regulatory subcategories (see Table 3-1).

The manufacturer must execute GEM for each truck sold and must comply with the standards using the averaging, banking, and trading tools offered in 40 CFR Part 1066 (of the Phase I Rule). If it is the case that the manufacturer offers sufficiently efficient vehicles and that the purchasers, for economic or external reasons, elect to purchase these sufficiently efficient vehicles, then the GEM computations may not be time-sensitive, because compliance is assured. However, GEM computations must be completed to facilitate annual federal reporting. If the manufacturer foresees that compliance may be assured only by incentivizing the purchase of more efficient vehicles over less efficient vehicles, the manufacturer may need to execute GEM for sales control purposes. This may be done with a sense of urgency, to maintain a record of running averaged efficiency values throughout the reporting year. Separate execution of GEM for each vehicle is impractical using manual entry, which was required by the locked version that is publicly available. OEMs have, therefore, developed automated techniques, including obtaining source code to allow better implementation within their order entry system.

Finding: For manufacturers with high sales volumes, the gathering of sales data and their efficient and rapid processing with GEM is of great importance to facilitate annual

TABLE 3-1 Fuel Consumption of Class 7 and 8 Vehicles by Regulatory Category and Subcategory (gallons per 1,000 miles)

		Regulatory Category		
		Day Cab		Sleeper Cab
Regulatory Subcategory and Effective Date	Type of Truck	Class 7	Class 8	Class 8
Mandatory				
MY2017 and later	Low roof	10.2	7.8	6.5
	Mid roof	11.3	8.4	7.2
	High roof	11.8	8.7	7.1
MY2016	Low roof	10.5	8.0	6.7
	Mid roof	11.7	8.7	7.4
	High roof	12.2	9.0	7.3
Voluntary				
MY 2013-2015	Low roof	10.5	8.0	6.7
	Mid roof	11.7	8.7	7.4
	High roof	12.2	9.0	7.3

SOURCE: 76 Fed. Reg. 57106-57513.

federal reporting. See Section "User interface, order entry, and GEM utility," below, for additional information.

ANALYSIS OF GEM

There is limited experience with the GEM model as the regulation is not yet in effect and only some OEMs have complied with the Phase I Rule, which takes effect in 2014 (Volvo Trucks, 2012; Daimler Trucks, North America, 2012; Reiskin, 2013). The validation of the model was limited to two vehicle tests and several computer runs against another simulation tool. Vehicle OEMs experienced with GEM have noted some errors that must be corrected.

It must be clear in considering GEM-based regulation whether the regulation is addressing national energy security, climate change, or economic benefit. This is of importance when alternative fuels are considered, and when any of these factors may be stronger influences than the regulation or run counter to the regulation. It would be optimal if GEM and the attendant rulemaking were capable of standing alone in driving the most appropriate technologies to raise fuel efficiency and if it did not rely on economic reality to ensure that inappropriate technology was not forced.

Future versions of GEM may need to take account of NHTSA's and the Federal Motor Carrier Safety Administration's (FMCSA's) plans to require speed limiters on all current and new vehicles. The exact value of the speed limit, and the techniques for dealing with smart speed limiters that provide additional speed for passing, are not yet known (Miller and Cama, 2013).

The comments provided on GEM in this section are in most cases inseparable from the regulation implied by GEM and the processes used to find constants for GEM. In what follows, some comments and findings may refer to the tool itself, others to the measurement, design, modeling, and regulation implied by the tool. With the changes being considered by EPA and the recommendations in this report, GEM has the potential to become recognized as the best model for predicting vehicle performance on routes.

User Interface, Order Entry, and GEM Utility

In general, the GEM user interface meets neither Recommendation 8-4 of *Assessment of Technologies and Approaches to Reducing the Fuel Consumption of Medium- and Heavy-Duty Vehicles* (NRC, 2010) nor the criteria established in Appendix G to that report. (These are discussed in the Section "Development of the Greenhouse Gas Emissions Model.") OEMs have requested the source code for the model in order to adapt it to their order entry system and deal with proving that the outputs of the model are consistent with the original version (EPA, 2011b, p. 7-5). The cost and time to implement the model into order entry systems are considerable.

Finding: The current GEM Version 2.0.1 is very much simplified, has a poor user input interface that is not compatible with the automated order entry systems of the vehicle OEMs, and has insufficient output information.

Recommendation 3.2: The GEM programmers should make every effort to configure GEM to be compatible with existing OEM order entry systems. For maximum effectiveness, a vehicle OEM should be able to easily run GEM on the initial specification of a product from a customer and on any changes that may ensue by customer choice, manufacturer choice, or supplier availability of parts.

Finding: A sufficiently accurate version of GEM could be used by OEMs and fleets for making significant trade-offs on technology purchases for vehicles and for benchmarking their operations.

Recommendation 3.3: GEM should be made to provide a more useful output that includes graphs and other presentation methods that will allow for greater insight into the actions that an OEM can take to improve GHG emissions and fuel efficiency. To this end, future versions of GEM must be sufficiently sophisticated to yield realistic predictions of truck efficiency and accurate predictions of efficiency changes in response to design variables for a variety of vehicle activities.

User-Specified Data Input and Hard-Coded Features of GEM

Weight and Rolling Resistance

The major GEM inputs are for rolling resistance of the tires and for drag coefficient of the vehicle. Generally, the truck manufacturer obtains the tire data from the tire manufacturer, but it must either conduct aerodynamic coast-down testing in accordance with 40 CFR Part 1066 or use an alternative acceptable method to obtain drag area.

The weight reduction input in GEM is limited to a fixed set of technologies and parts. This does not successfully represent the efforts of OEMs to search for weight savings in all new part designs and might discourage innovation. The goal is to minimize the weight of the vehicle and maximize the weight of the load to achieve the optimum freight efficiency as measured in ton-miles while still meeting all required road, load, bridge, and safety requirements. Weight reduction credit might better be focused on achieving a total vehicle weight less than a given number.

In contrast to the well-established process for determining engine efficiency and GHG emissions, the GEM inputs for tire rolling resistance and aerodynamic drag are obtained using processes that were defined more recently and for which the variability is less well documented.[5] Obtaining tire rolling resistance values was initially reported to be a problem for OEMs (as the committee learned through presentations by Ford,[6] GM,[7] and Navistar[8]). Calibration of equipment traceable to national standards for passenger car tires is in place in Europe and self-calibration of equipment is in place for the grading of tires. Calibration of truck tire characterization equipment needs to be implemented in U.S. regulations.

Finding: For tire rolling resistance, there needs to be high confidence in input values to GEM, because fuel efficiency improvement rests on incremental changes in technology performance. Accuracy, as well as agreement between repeat tests and different laboratories, needs to be sufficiently good to permit detecting changes in performance.

Recommendation 3.4: A mechanism needs to be implemented for obtaining accurate tire rolling-resistance factors, including equipment calibration, and for maintaining that information in a public database. This might be managed in the same way that tread wear, temperature, and traction data are displayed through the federal Uniform Tire Quality Grading system.

Much has been learned about measuring aerodynamic performance since the NRC Phase One Report was published and the Phase I Rule promulgated. There needs to be high confidence in input values to GEM, because fuel efficiency improvement rests on incremental changes in technology performance.

Recommendation 3.5: Regulators should refine and improve the methods for obtaining aerodynamic performance data that better reflect real-world experience, including yaw and varying speeds.

Hard-Coded Features of GEM

GEM specifically does not allow for synergy between components, the operation or control of components in a most efficient way, or the engendering of efficiency through operation of a smaller component at higher relative load. In this way some opportunities for reduction in fuel use are lost. GEM specifies the performance maps for major components such as the engine and transmission but does not credit the vehicle manufacturer with benefits of using a potentially superior engine or transmission. Several presenters have suggested that full vehicle simulation or simulation of a power pack that includes the aftertreatment system and the transmission should be used, rather than only the engine. These approaches are intended to show the benefit of optimizing system-level operation rather than component-level operation. The idea of power pack versus engine only can apply either to test cells or to simulation. In this interim report, the committee supports the extension of current certification to include a power pack in a next-generation GEM simulation by the OEM. As indicated in the introduction to this chapter, the committee will consider full-vehicle simulation and other approaches to engine certification in the final report.

GEM is, accordingly, too simple to have direct value in the manufacturer's vehicle design process. Rather, it influences design by virtue of its required use in demonstrating and achieving compliance. At this time of writing, the economic motivator of high fuel cost has as much influence as GEM on design, but it is inappropriate to rely on an economic argument to keep GEM from driving possibly unproductive outcomes.

Finding: In having limited inputs, in treating those inputs in a simple fashion, and in not allowing for skillful design integration, GEM (both the model and the regulation implied in using the model) may encourage designs that are suboptimal.

GEM employs a limited set of cycles to challenge the simulated truck. These cycles do not include real-world road grade. Being speed and time based, these cycles also do not allow for the faster acceleration of more powerful trucks, or the longer times that might be taken by less powerful trucks to complete some real-world routes. This deficiency is not evident when a hard-coded engine map, rather than a real OEM engine simulation, is used in the model.

Finding: Critical issues in the choice of test cycles in GEM include road-with-grade and different speed and torque profiles to better relate the test or model to real-world experience. In addition, route concepts or distance-based target schedules might provide a superior alternative to speed and time cycles.

Recommendation 3.6: The choice of test cycles and routes or schedules used in GEM needs to be readdressed thoroughly to avoid creating designs that are optimized for the

[5] EPA has indicated that it has a contract with SwRI for "conducting SAE J1321 (fuel consumption benefit) testing on the road using verified tires." Sam Waltzer and Cheryl Bynum, EPA, "SmartWay Technology Program: Influencing Efficient Freight Movement into the Future," Personal communication to Tom Cackette and Chuck Salter, NRC Committee on Technologies and Approaches to Reducing the Fuel Consumption of Medium- and Heavy-Duty Vehicles, Phase Two, July 24, 2013.

[6] Ken McAlinden, Ford Motor Company, "Heavy Duty GHG from a Full-Line Manufacturer's Perspective." Presentation to the committee, June 20, 2013.

[7] Mark A. Allen and Barbara Kiss, "General Motors comments: NAS Panel on Heavy-duty GHG/CAFE Discussion," Presentation to the committee, July 31, 2013.

[8] Greg Fadler, Navistar, Inc., "Navistar Fuel Economy and Emissions," Presentation to the committee, March 20, 2013.

test rather than for achieving real-world performance in the design process.

If OEMs were allowed to substitute any models or real test data for parts of GEM at will, a continuum for regulation would evolve between present-day GEM certification and whole-vehicle testing. This could be addressed by causing GEM to offer current practice default models for components, and, if a manufacturer's component can exceed the performance of that default model, the model for the actual component could be substituted. Full vehicle modeling might also be addressed in the final report. Models should be capable of simulating real-world component behavior accurately and should not be oversimplified.

Recommendation 3.7: NHTSA should investigate allowing the OEM to substitute OEM-specific models or code for the fixed models in the current GEM, including substituting a power pack (the engine, aftertreatment, transmission). These models, whether provided by OEMs or fixed in the code, should be configured to reflect real-world operation accurately.

Fixed Values in the GEM Code

Certain parameters whose values can vary significantly in actual practice are fixed in GEM. As noted, GEM specifies the performance maps for major components such as the engine and transmission and does not credit the manufacturer with benefits for using a potentially superior engine or transmission, singly or in combination, such as occurs with predictive cruise control. Engines are certified separately in a test cell and measured at several points over specified drive cycles. In contrast, GEM addresses a steady-state condition for the speed and load of the vehicle. Therefore, the results may not represent the best fuel consumption over drive cycles that are expected to include different road speeds, grades, loads, and yaw angles.

Finding: GEM output is unaffected by the actual use of a smaller or larger engine in a truck in the same subcategory, because the engine map used by GEM is predefined. For example, downsizing of the engine is a known approach for saving fuel that GEM does not properly address.

Recommendation 3.8: The regulators should assess whether a steady-state speed-torque map is sufficient for GEM accuracy in engine efficiency prediction.

Similarly, GEM, in having a fixed test weight, does not differentiate between lighter duty Class 8 tractors that may pull volume-limited loads and heavier duty tractors that typically pull combined gross vehicle weights approaching 80,000 lb. Although single-axle sleeper cabs may offer an opportunity to reduce fuel use for volume-limited loads, they are not a subcategory in GEM.

Finding: The use of measured values can allow optimization of the engine/transmission/driveline (such as a power pack) to show its positive impact on the environment.

Recommendation 3.9: NHTSA, in coordination with EPA, should investigate substituting measured values for the fixed vehicle weights in the current model.

Vehicle and Component Integration

Controls, Vehicle Integration, and the Role of Aftertreatment

GEM and the implementation of GEM must encourage and facilitate all or most technology avenues that offer significant reduction in fuel consumption, by allowing accurate, quantitative modeling that does not restrict any fruitful options. Fuel consumption reductions that involve vehicle integration, power train integration, control strategies, auxiliary engine loads, aftertreatment strategies, or integrated power packs must be encompassed.

Finding: There are many technology avenues for reduction of fuel consumption that are currently not captured by GEM.

Recommendation 3.10: NHTSA should consider carefully ways in which a revised GEM simulation and associated test procedures can reflect the benefit of integrating an engine, aftertreatment, and transmission with interactive controls. NHTSA should consider whether a power train, or a power pack consisting of engine, aftertreatment, and transmission, can be certified for fuel use and GHG emissions in the manner of an engine; if that path is adopted, the power:weight ratio of the vehicle must be considered equitably.

Real-world operations and routes need to be considered for use in GEM, and fleet data could be used to develop or validate the target activity in GEM. Another source of information is the Federal Highway Administration Freight Performance Measures Initiative. As a guiding principle, operators of vehicles need to move freight safely and efficiently or otherwise perform work with a vocational vehicle. If properly designed, regulations and their implementation in GEM would not compel equipment that is unsuited to real-world operation, nor would GEM deprive a purchaser of performance features that are truly necessary for service but that may be atypical of current GEM cycles. For example, engine downsizing or lighter, lower friction, less robust transmissions might yield fuel savings but might not be appropriate for certain duty cycles.

The current supplementary emissions test cycle based on 65 mph and a 38,000 lb load might be good for providing a certification cycle that is simple, but it does not reflect the real-world operation of over-the-road tractors that, according to reports from fleets, average less than 50 mph when taking into account regional operation and traffic. GEM should be capable of dealing equitably with, for example, a truck that cannot follow a speed-time trace owing to insufficient power over a part of the cycle or a truck so powerful that the cycle does not approach full power use.

Finding: GEM does not satisfactorily reflect real-world operation for over-the-road tractors.

Recommendation 3.11: NHTSA and EPA should modify the Greenhouse Gas Emissions Model (GEM) to employ cycles or vehicle activities that cover as large a fraction of over-the-road tractor operation as possible without becoming overly cumbersome. GEM should employ a sufficient number of truck types or subcategories to facilitate sound and beneficial regulation.

Tractor and Trailer

The tractor and trailer are fundamentally inseparable in addressing aerodynamic drag and design, but should also consider that present-day tractor and trailer fleets may limit realization of the benefits of integrated tractor and trailer design. If trailers are to be included in a future rule, GEM may need to be modified in some way to account for the interaction between the tractor and trailer. While a rule can be set to deal with the trailer alone, or a standardized trailer can be employed for aerodynamic testing, there are significant aerodynamic issues associated with the tractor-trailer gap and the air flow from under the tractor to under the trailer. NHTSA and EPA must avoid separate optimization of components, as this might create system-level issues and may prove counterproductive to true optimization. However, if the default is that a standard trailer must be used, that trailer must reflect best practice technology for efficiency improvement.

Finding: Using GEM to address all tractors in combination with trailers would accrue two principal benefits. First, low-rolling-resistance tires and aerodynamic devices are likely to offer benefits with tankers or flatbed trailers. Second, the current bobtail testing (i.e., operating a tractor without a trailer attached) of certain tractors is unlikely to faithfully represent their performance in combination with trailers.

Recommendation 3.12: NHTSA should assess the benefit of using GEM to address all tractors in combination with trailers.

Inclusion in GEM of Engines Using Alternative Fuels

Since natural-gas-powered vehicles are expected to become significantly more prevalent in the coming years, regulatory changes will likely aim at modifying GEM to incorporate these vehicles rather than driving them toward a separate compliance path. It would be preferable to use a model for the actual engine in GEM. Otherwise, the current procedures for advanced technology credits may be replaced by allowing the inclusion of a natural gas engine model in addition to a diesel engine model. (Natural-gas-fueled vehicles and the possible regulatory approaches to them are discussed in Chapter 5.) In a similar fashion, if any other fuel or combination of fuels shows a likelihood of substantial market penetration by 2017, it should be considered and modeled in the power train.

Information on Fuel Consumption

Other methods of obtaining fuel economy information could be investigated, such as making use of the following: any onboard computers required in the future by FMCSA; reliable reporting information already available to the government; and industry statistics that are regularly gathered, such as ton-miles and fuel taxes. Cooperation of major fleets in assessing real-world fuel economy would provide valuable validation or correction for GEM and raise public confidence in the fuel efficiency regulations. SmartWay fleets are required to provide data on an annual basis that could be of benefit.

On a monthly basis, each state is required to report to the Federal Highway Administration (FHWA) the number of gallons taxed by that state. These data are analyzed and compiled by FHWA staff. The data on the amount of on-highway fuel use for each state is then used to apportion federal revenue to each state. Yearly, the FHWA's Office of Policy provides the previous year's data for use in the attribution process. This allows the states extra time to review the data and verify that it is correct and ready to be used in attribution.[9]

REFERENCES

Daimler Trucks. 2012. "North America: Daimler Trucks North America leads industry by certifying complete vehicle lineup GHG14 compliant," http://media.daimler.com/dcmedia/0-921-657777-1-1466206-1-0-1-0-0-0-11701-1549054-0-1-0-0-0-0-0.html. Accessed November 15, 2013.

Environmental Protection Agency (EPA). 2011a. *Peer Review of the Greenhouse Gas Emissions Model (GEM) and EPA's Response to Comments*. EPA-420-R-11-007. Washington, D.C.: EPA. August.

EPA. 2011b. *Greenhouse Gas Emissions Standards and Fuel Efficiency Standards for Medium- and Heavy-Duty Engines and Vehicles: EPA Response to Comments Document for Joint Rulemaking.*

EPA and National Highway Traffic Administration (NHTSA). 2011. *Final Rulemaking to Establish Greenhouse Gas Emissions Standards and Fuel Efficiency Standards for Medium- and Heavy-Duty Engines and Vehicles Regulatory Impact Analysis.* EPA-420-R-11-901. Washington, D.C.: EPA. August.

[9] See http://www.fhwa.dot.gov/policyinformation/motorfuelhwy_trustfund.cfm. Accessed November 15, 2013.

Miller, E., and T. Cama. 2013. DOT's speed-limiter proposal to add older trucks, aide says. *Transport Topics*. November 4.

National Research Council (NRC). 2010. *Technologies and Approaches to Reducing the Fuel Consumption of Medium- and Heavy-Duty Vehicles.* Washington, D.C.: The National Academies Press.

Reiskin, J.S. 2013. Navistar, Paccar continuing efforts to achieve 2014 GHG certification. *Transport Topics*. November 11.

Volvo Trucks. 2012. "Volvo trucks earn 2014 greenhouse gas certification for entire Class 8 vehicle lineup." www.volvotrucks.com/trucks/na/en-us/_layouts/CWP.Internet.VolvoCom/Newsitem.aspx?News.Itemid=135785&News.Language=en=gb. Accessed November 15, 2013.

4

Baseline Information on MHDV Fleet and Methodology for Collection

INTRODUCTION

The committee was assigned the following task regarding baseline information:

> The committee will review an updated analysis of the makeup and characterization of the medium- and heavy-duty truck fleet, including combination tractors, trailers, busses and vocational vehicles. The committee also will review the methodology for providing on-road information on fuel consumption.

The committee's work is thus similar to Task 3, "Model Year 2011+ Fleet Baseline," as set forth in the National Highway Traffic Safety Administration (NHTSA) contract with Southwest Research Institute (SwRI).[1,2] The NHTSA objective for the contract with SwRI is

> Augment information available on baseline for the MY2011+ fleet by vehicle segment (engines compliant with 2010 criteria pollutant requirements).

The scope is

> Statistics on 2011+ MD/HD on-road fuel consumption and emissions by vehicle class, information on components and technology penetration, vehicle types and duty cycles within each class, how many Class 2b and 3 vehicles are chassis vs. engine dyno certified, share of empty miles within each class and share of alternate fuel vehicles.

The schedule is

> In process of procuring all available R.L. Polk registration data and forecasts for Class 2b-8 vehicles (including motor coaches, RV's and trailers).
>
> Fleet Manager Survey: Frost & Sullivan to survey a minimum of 100 motor carrier fleet managers for in-use characteristics of latest regulatory generation of Class 2b-8 vehicles including fleet make-up (e.g., final vehicle builds and vocational), typical use (mileage, fuel economy and freight tonnage), and fuel economy/emissions technology levels.

Since little of the requested information has been received from NHTSA, SwRI, R.L. Polk, or Frost & Sullivan, the committee has focused its efforts on addressing NHTSA's procedure for establishing a useful baseline. This chapter covers the following: (1) why do we need a baseline?; (2) criteria for a good baseline collection process; (3) comments on NHTSA, SwRI, and Frost & Sullivan survey; and (4) comments on the CalHEAT report for the California Energy Commission. It also contains Findings and Recommendations, as well as annexes.

WHY DO WE NEED A BASELINE? NHTSA SHOULD HAVE A BASELINE IN ORDER TO INFORM ITS RULEMAKING

What Is a Baseline?

A baseline can measure different characteristics of the truck fleet, depending upon the ultimate use of the information. In regulatory proceedings, a baseline may refer to a baseline vehicle in each regulated category for the purpose of applying a regulatory standard. More broadly, a baseline may characterize the truck fleet at a given point in time with respect to types of vehicles in use, technologies being deployed (for example, technologies such as auxiliary engine loads and aftertreatments were left out of GEM's first iteration) on vehicles in each vehicle class, work performed by the vehicles, and fuel consumption of individual vehicles as well

[1] James Tamm, National Highway Traffic Safety Administration, "NHTSA MD/HD Vehicle Fuel Efficiency Technology Study," Personal communication to NRC staff, April 8, 2013.

[2] Tom Reinhart, Southwest Research Institute Labs, Presentation to the committee on June 25, 2013.

as the fleet as a whole. It is in the latter sense that NHTSA has asked the committee to examine the data available and devise a methodology for collecting data to establish a baseline.

Why a Baseline?

It is important to measure how the industry responds to regulation so that subsequent phases of regulation can be based on a clear view of the impact of previous regulatory stages. Comparing subsequent years' data to the baseline data will help to measure the impact of regulations. For example, one of the clear risks of regulating such a complex industry as trucking is the risk of unintended consequences. Tracking change over time will allow NHTSA to observe unintended consequences and allow correction in subsequent phases of regulation. Monitoring performance over time will also allow NHTSA to observe and measure obstacles to technology adoption, which can improve the effectiveness of subsequent stages of regulation. Knowledge of technology penetrations in the baseline will help guide the stringency of future standards with respect to the effects of projected penetration trends.

One could also use baseline data to measure how the fleet would change in response to key external developments such as fuel price changes or technology developments in the absence of regulation. The large original equipment manufacturers (OEMs) have simulation-based specification-defining tools that seek to give the customer an efficient vehicle for meeting freight and work requirements. Regulations, by mandating equipment and simulation tools, may force vehicle choices that differ from these OEM efforts to provide efficient vehicles. Using the OEM simulation tools one could calculate changes from the baseline that would have taken place absent the regulations. In this way, a potentially useful measure of the effects of regulation could be obtained.

Which Year Should the Baseline Capture?

Since regulations are set for new vehicles, the baseline should be representative of recent vehicle purchases and recently promulgated regulations. In that way, NHTSA will be able to analyze *incremental* effects of new regulations. In particular, the baseline should capture the effects of the 2010 NO_x standards. In addition, it would be desirable to have more complete information on the effects of the Phase I Rule on the fleet before setting Phase II standards. However, the schedule NHTSA has established for data gathering and analysis seems to preclude a baseline year that largely incorporates the effects of 2014 greenhouse gas (GHG) standards. Consequently, model years (MY) 2010, 2011, 2012, and 2013 would be the most appropriate years for establishing the baseline. It can be seen from the press releases of various manufacturers that some portions of the 2013 product mix have already been certified to the 2014 standards. Those data provide a small window into manufacturer compliance.

Which Data Should the Baseline Contain?

Ideally, the data should describe the features of the fleet relevant to rulemaking. The information would include:

- *Vehicle data.* It is important to collect information on a vehicle basis. While important for understanding national trends, fuel consumption and ton-miles driven on a fleet basis will not sufficiently inform the setting of regulations. Consequently, the database should collect information on a vehicle basis, including the type of vehicle (tractor-trailer, straight truck, motorcoach), the engine type (diesel, natural gas, gasoline, hybrid), the fuel-saving technologies (aerodynamics, tires, transmissions, axles), ton-miles driven, work performed, and fuel consumption. This will allow NHTSA to measure the adoption rates of specific technologies, monitor their performance, and assess market barriers to adoption. The individual vehicle data can then be aggregated by segment where necessary to arrive at updated parameters for use in modeling and simulation.
- *In-use data.* Collection of data on real-world experience will allow NHTSA to update the GHG Emissions Model (GEM) as well as to perform more accurate cost-benefit analyses of energy-saving technologies. Ideally, the database will reflect real-world experience on the use of fuel-saving technologies and include information that quantifies the experience with fuel-saving technologies, including estimated returns on investment.
- *Sample.* Data are needed on small as well as large trucking fleets. The trucking industry is very diverse, with many small fleets accounting for a significant proportion of total trucks in use. This diversity resides in the size of the fleet, whether the user owns or leases equipment, the types of operation of the vehicles (e.g., freight movement, services such as ambulances or utility bucket trucks, construction, or refuse hauling), and whether the operation is for hire or private carriage. Combinations of these attributes exist in most all segments, and the baseline should either include them all or specifically state which ones are not being included. It appears that the R.L. Polk data will be procured for Classes 2b through 8 vehicles, including motor coaches, recreational vehicles (RVs), and trailers, but the Frost & Sullivan survey is for 100 motor carrier fleet managers. It is not clear how NHTSA will gather in-use characteristics for nonfleet vehicles such as personal pickups and vans, personal motor coaches, RVs, and other nonfleet vehicles.

CRITERIA FOR A GOOD BASELINE DATA COLLECTION PROCESS

The data collection process should be replicable so that NHTSA will be able over time to measure the impacts of regulation, improve the efficiency of regulation, and adjust for unintended consequences. For the data to be replicable, the collection process should not be too large a burden on respondents in terms of cost and time. A data collection process that is too burdensome is likely to be ignored or result in spotty responses that would undermine the quality of the data.

COMMENTS ON NHTSA, SWRI, AND FROST & SULLIVAN SURVEY

On December 19, 2012, NHTSA published in the Federal Register[3] a proposal for collection of information on Medium- and Heavy-Duty Trucks. Comments were received from several parties based on the following draft publications:[4]

- Frost & Sullivan Fleet Profile Screener Draft v4 December 4, 2012, and
- Questionnaire Draft v8 December 4, 2012.

No subsequent drafts were made available to the committee for review. Based on the committee's review of the drafts received and comments made on the drafts by the parties that provided comments on the federal notice[5] (American Trucking Associations, Daimler Trucks North America, NAFA Fleet Management Association, National Transportation Equipment Association, Recreational Vehicle Industry Association, and the Truck and Engine Manufacturers Association), the committee judges that the survey approach and questions asked are problematic and unlikely to provide the needed information. In short, the survey has the following shortcomings:

- Limited to fleets of 100 or more vehicles, which does not capture the diversity of the trucking companies that make up the fleet.
- Limited to 100 respondents.
- Focused heavily on Class 8 on-road fleets and does not represent vocational fleets.
- Unclear with respect to model year designation, because engine MY and vehicle MY might differ.
- Load information is difficult to assess.
- Trailer and work truck body equipment is limited and too basic.
- Information on technologies utilized is limited and does not include additional fuel-saving practices of fleets.
- The survey will take much longer to complete than suggested. Finding the information requested will be difficult and burdensome.
- The survey does not take into account tractors that are used for both long-haul and local delivery.
- Information is aggregated and not broken down into class or vocation.
- Engine and transmission information is limited, along with information on other major components.
- Some OEMs have begun to manufacture products that already meet GHG 2014 regulations.

A number of commenters noted the focus on 2010 and newer engines and vehicles. The reason for this focus and the purpose of the survey were not clear to some commenters. The committee understands the purpose of the survey is to obtain data for creating a baseline of information prior to the implementation of the current 2014 GHG regulations for MHDVs. The committee also believes that information will be needed as the regulations take effect, and consideration is given to future regulations. Since NHTSA and the Environmental Protection Agency (EPA) concentrate on the manufacture of new vehicles in these regulations, the committee understands the need to instead focus only on vehicles that meet the current 2010 engine emissions regulations. These would be vehicles with MY2010 engines, regardless of the vehicle model year.

NHTSA should employ alternative means to develop a reliable and repeatable process for obtaining the information that is needed to inform the efficacy of current regulations and promulgation of future regulations.

COMMENTS ON THE CALHEAT REPORT FOR THE CALIFORNIA ENERGY COMMISSION

The California Hybrid, Efficient and Advanced Truck Research Center (CalHEAT) recently completed a study of trucks in California (Silver and Brotherton, 2013). The report provides valuable information related to the baseline of vehicles in California. The methodology used may serve as a model for developing a baseline for commercial vehicles in the entire United States. The abstract from that study reads as follows:

> The California Hybrid, Efficient and Advanced Truck Research Center (CalHEAT) was established by the California Energy Commission in 2010. It is operated by CALSTART to perform research into planning, commercializing, and demonstrating truck technologies for more fuel-efficient medium- and heavy-duty vehicles and to reduce emissions. The role of the research center is to coordinate the development of a *Research and Market Transformation Roadmap* to deliver clear actionable steps to help meet or exceed the

[3] 77 Fed. Reg. 75257.

[4] Available at http://www.regulations.gov/#!docketDetail;D=NHTSA-2012-0170. Accessed September 30, 2013.

[5] See http://www.regulations.gov/#!docketDetail;D=NHTSA-2012-0170. Accessed September 30, 2013.

2020 goals for California in petroleum reduction, carbon reduction, and air quality standards, and identify longer term goals through 2050. Medium- and heavy-duty trucks account for 9% of greenhouse gases in California, and approximately 20% of fuel consumption. Improvements in efficiency or reduction of petroleum use by trucks provide a substantial opportunity to reduce emissions. (Silver and Brotherton, 2013)

The CalHEAT reports also note that

> . . . as the first step in the development of this Roadmap, CalHEAT performed a California Truck Inventory Study to better understand the various types of trucks used in California, their relative populations, and how they are used. The analysis included nearly 1.5 million commercial medium- and heavy-duty trucks, grouped by weight and application, to establish a baseline inventory and determine fuel use and potential for efficiency and emissions improvements. CalHEAT also conducted Phase I research to characterize the California truck population by size, use, and emissions, and prepared a baseline report of available technology and pathways for improvement. Phase II research identified gaps along the pathways and barriers to progress, and developed a decision-making tool to identify the most efficient choices to meet the State's goals. (Silver and Brotherton, 2013, p. 61)

FINDINGS AND RECOMMENDATIONS

The committee recommends as follows:

Recommendation 4.1: NHTSA and EPA should implement the following practices as part of a good baseline data collection process:

- To the extent possible, existing data sources should be exploited. NHTSA should take advantage of existing private sources such as R.L. Polk. (See the annex to this chapter for other sources of baseline data.)
- Data already collected in connection with certification of medium- and heavy-duty vehicles (MHDVs) should be utilized in establishing the baseline. To the extent other agencies collect data on MHDVs, the data collection should be harmonized and streamlined to minimize the collection burden.
- Other extant data analysis processes in the industry (e.g., data products of firms that interview truck builders and fleets for industry forecasting) should be utilized. Here, the processes could be expanded to satisfy NHTSA's ongoing data collection needs.

Finding: There is no existing baseline of data that meets the needs of a baseline for the MHDV fleet. The Frost & Sullivan survey process, as presented to the committee in documents dated December 2012, will not provide reliable and repeatable information.

Finding: The collection of R.L. Polk registration data and forecasts is appropriate for Class 2B to Class 8 vehicles (including motor coaches, RVs, and trailers). It is the recognized source within the industry.

Recommendation 4.2: NHTSA should establish a repeatable, reliable data collection process as soon as possible. In addition to continuing data procurement with SwRI and R.L. Polk, NHTSA should investigate outside sources—such as FTR, ACT Research, SmartWay, the North American Council for Freight Efficiency (NACFE), and the American Transportation Research Institute (ATRI)—to obtain a repeatable, reliable baseline as well as future data. These sources could use the R.L. Polk data and conduct clarifying, deeper interviews with truck and trailer builders, manufacturers, and fleets to elicit specific ongoing data on technologies procured and fuel consumption.

Recommendation 4.3: Because the government operates a wide variety of trucks and buses that are a potential source of data, NHTSA should explore with the General Services Administration, the U. S. Postal Service, and the Federal Transit Administration (FTA) the advisability of building a database on the government-operated truck fleet and municipal bus fleets purchased with grants from FTA. The database should capture information on new vehicle purchases by model year, fuel-saving technologies utilized, maintenance and operating costs, and the fuel-consumption performance of the vehicle.

Finding: The CalHEAT study produced a valuable inventory or baseline for the State of California using R.L. Polk data. In addition, the data from the discontinued Vehicle Inventory and Use Survey (VIUS) was useful in the past for safety analysis, freight planning, and transportation system analysis.

Recommendation 4.4: NHTSA should establish a repeatable survey for private fleets. Possible models for such a survey include the CalHEAT study, the discontinued VIUS methodology, and other private data sources detailed in the annex to this chapter.

REFERENCES

American Trucking Associations (ATA). 2012a. *U.S. Freight Transportation Forecast to 2023.*

ATA. 2012b. *American Trucking Trends: 2012.*

Silver, F., and T. Brotherton. 2013. *Research and Market Transformation Roadmap to 2020 for Medium- and Heavy-Duty Trucks.* CEC-XXX-2013-XXX. Draft Rev # 7 Dated 6-14-2013. Sacramento, Calif.: California Energy Commission.

ANNEX 4A
OTHER SOURCES OF BASELINE DATA IN THE INDUSTRY

Because many aspects of commercial vehicles and hauling of freight are regulated by various government entities, information is often publicly collected, aggregated, and disseminated to the industry. Some is done by government agencies, such as the U.S. Department of Energy and the U.S. Department of Transportation. Some is done by trade organizations that help their members do benchmarking against competitors to improve their operations. The following seven sources will be discussed briefly; however, this is not an exhaustive list for creating a baseline for future regulations or for creating a new industry metric tracking or estimating GHG emissions, fuel economy, fuel efficiency, or freight efficiency.

- FTR
- ACT Research
- American Trucking Associations (ATA) reports
- ATRI
- Wards Auto
- R.L. Polk
- North American Council for Freight Efficiency (NACFE)

FTR

FTR is one of two organizations that are used monthly by companies throughout trucking to understand the business of hauling freight.[6] Using both public and private sources, including regular interviews with fleets, it compiles information ranging from current registrations of vehicles through industry trends to freight hauled. Annex Figures 4A-1 through 4A-4 and Tables 4A-1 and 4A-2 exemplify just a few of the information displays from a typical report. This organization should be contacted to determine if the needed

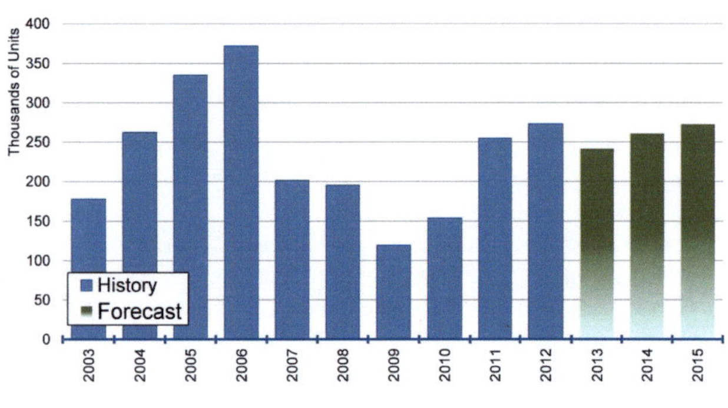

ANNEX FIGURE 4A-1 North American Class 8 Factory Shipments (annual). SOURCE: FTR.

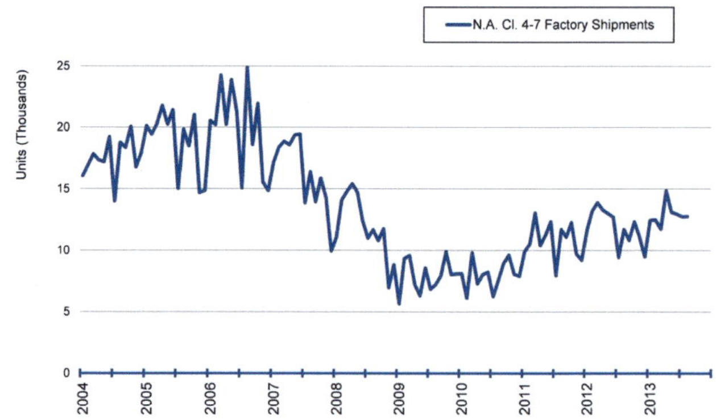

ANNEX FIGURE 4A-2 North American Class 4 through Class 7 Factory Shipments. SOURCE: FTR.

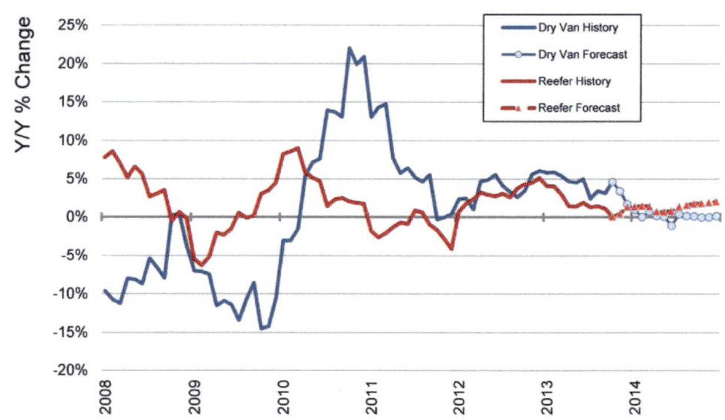

ANNEX FIGURE 4A-3 Truck loadings: dry van and refrigerated trailers. SOURCE: FTR.

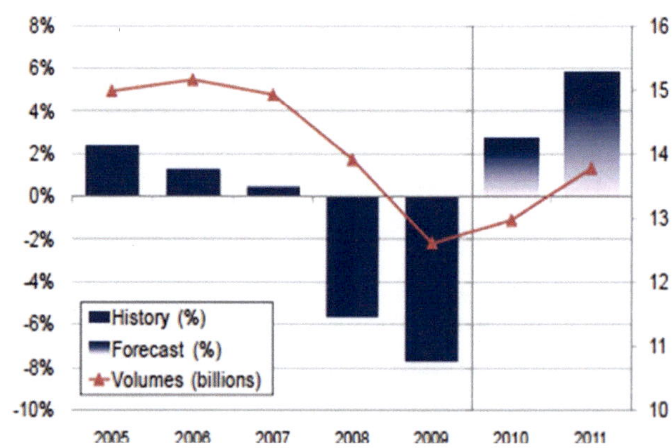

ANNEX FIGURE 4A-4 U.S. truck ton-miles (total). SOURCE: FTR, copyright 2010.

[6] Further information is available at http://www.ftrassociates.com.

ANNEX TABLE 4A-1 Quantities of Freight Hauled in Class 8 Vehicles, Historic and Projected

	2013				2014				Annual			
	Q1	Q2	Q3	Q4	Q1	Q2	Q3	Q4	2012	2013	2014	2015
Tons (000,000)	2,917	2,942	2,982	3,009	3,021	3,035	3,045	3,083	11,334	11,850	12,185	12,458
Y/Y change (%)	3.60	4.10	5.80	4.80	3.60	3.20	2.10	2.50	2.20	4.60	2.80	2.20
Ton-miles (000,000)	652,305	658,222	660,476	667,090	667,971	670,206	670,737	681,183	2,518,788	2,638,092	2,690,097	2,741,162
Y/Y change (%)	5.10	4.80	4.50	4.60	2.40	1.80	1.60	2.10	1.30	4.70	2.00	1.90
Length of haul												
Short haul	87,653	88,676	90,419	90,788	91,526	92,343	92,937	93,505	345,139	357,536	370,312	380,489
Y/Y change (%)	1.80	2.90	5.40	4.20	4.40	4.10	2.80	3.00	2.30	3.60	3.60	2.70
Medium haul	323,425	322,483	324,364	327,478	328,223	329,402	329,258	335,740	1,233,436	1,297,751	1,322,623	1,346,680
Y/Y change(%)	5.90	5.00	5.10	4.80	1.50	2.10	1.50	2.50	0.70	5.20	1.90	1.80
Long haul	241,227	247,063	245,693	248,824	248,221	248,461	248,543	251,938	940,213	982,806	997,163	1,013,994
Y/Y change (%)	5.2	5.30	3.30	4.30	2.90	0.60	1.20	1.30	1.90	4.50	1.50	1.70
Type of good												
Raw	90,839	92,554	94,951	96,814	96,632	96,891	96,298	99,603	354,931	375,158	389,423	394,503
Y/Y change (%)	2.20	4.60	8.20	7.80	6.40	4.70	1.40	2.90	0.20	5.70	3.80	1.30
Processed	226,466	229,690	231,407	232,988	233,968	234,944	236,030	238,557	874,191	920,551	943,500	967,374
Y/Y change (%)	5.10	5.50	5.00	5.60	3.30	2.30	2.00	2.40	0.70	5.30	2.50	2.50
Manufactured	334,999	335,978	334,118	337,288	337,370	338,371	338,409	343,023	1,289,667	1,342,383	1,357,173	1,379,284
Y/Y change (%)	5.90	4.50	3.10	2.90	0.70	0.70	1.30	1.70	2.10	4.10	1.10	1.60

NOTE: Data are seasonally adjusted. Y/Y, year to year.
SOURCE: FTR.

ANNEX TABLE 4A-2 Quantities of Freight Hauled in Classes 5 Through 7 Vehicles, Historic and Projected

	2013				2014				Annual			
	Q1	Q2	Q3	Q4	Q1	Q2	Q3	Q4	2012	2013	2014	2015
Tons (000,000)												
Total Classes 5-7	362	362	365	368	370	372	373	378	1,400	1,457	1,494	1,524
Y/Y change (%)	3.90	3.60	5.1	3.70	2.10	3.00	2.20	2.80	2.00	4.10	2.50	2.00
Ton-miles (000,000)												
Total Classes 5-7	33,874	33,887	33,942	34,214	34,291	34,382	34,374	34,918	130,884	135,917	137,966	140,134
Y/Y change (%)	4.70	4.00	3.40	3.20	1.20	1.50	1.30	2.10	1.10	3.80	1.50	1.60
Length of haul												
Short haul	23,214	23,039	23,176	23,341	23,445	23,566	23,576	23,963	89,140	92,770	94,550	96,236
Y/Y change (%)	5.00	3.90	4.00	3.40	1.00	2.30	1.70	2.70	0.80	4.10	1.90	1.80
Medium haul	9,939	10,114	10,035	10,142	10,138	10,136	10,125	10,291	38,886	40,230	40,691	41,288
Y/Y change (%)	4.20	4.30	2.40	2.90	2.00	0.20	0.90	1.50	1.40	3.50	1.10	1.50
Long haul	722	734	730	731	708	680	673	664	2,858	2,917	2,726	2,610
Y/Y change (%)	1.50	3.70	1.60	1.50	−1.90	−7.40	−7.80	−9.20	2.30	2.10	−6.60	−4.20
Type of good												
Raw	4,572	4,638	4,767	4,844	4,847	4,859	4,834	4,978	17,974	18,821	19,517	19,741
Y/Y change (%)	1.20	3.40	7.40	6.80	6.00	4.80	1.40	2.80	0.10	4.70	3.70	1.10
Processed	8,317	8,380	8,332	8,376	8,409	8,443	8,470	8,574	32,128	33,405	33,897	34,620
Y/Y change (%)	4.80	4.50	2.90	3.60	1.10	0.80	1.70	2.40	1.20	4.00	1.50	2.10
Manufactured	20,986	20,870	20,842	20,994	21,036	21,080	21,071	21,366	80,781	83,692	84,553	85,772
Y/Y change (%)	5.50	3.90	2.80	2.30	0.20	1.00	1.10	1.80	1.20	3.60	1.00	1.4

NOTE: Data are seasonally adjusted.
SOURCE: FTR.

information is already available, or if it can be obtained. Also, going forward, with this level of information it may be possible to create new metrics related to GHG emissions, fuel economy, fuel efficiency, or freight efficiency.

ACT Research

ACT Research is a second organization that publishes reports monthly used by companies throughout trucking. Its website is http://www.actresearch.net. Using both public and private sources, it compiles information on tonnage, vehicle population, and vehicle unit age, among others. Annex Figures 4A-5 through 4A-11 show a few sample graphs from a typical monthly report. This organization should be contacted to determine if the needed information is already available and accessible.

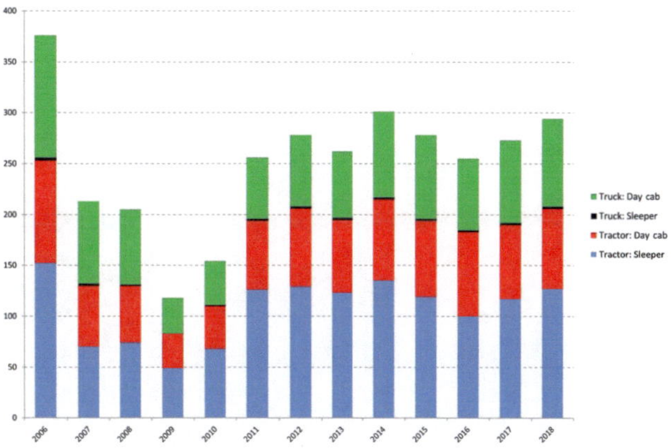

ANNEX FIGURE 4A-7 North American Class 8 production. SOURCE: ACT Research.

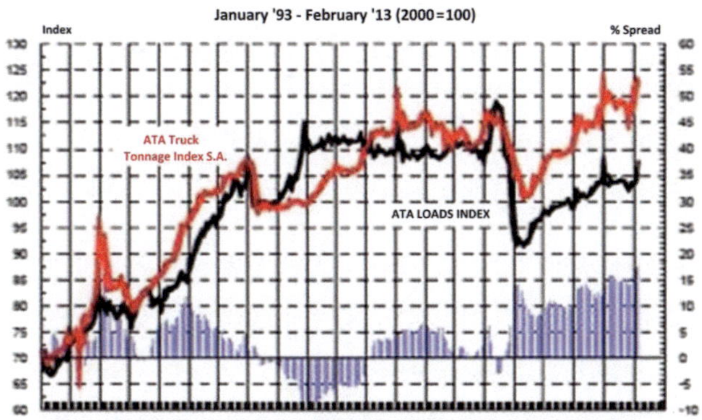

ANNEX FIGURE 4A-5 ATA truck tonnage index and ATA truck loads index NOTE: S.A., seasonally adjusted. SOURCE: ACT Research, copyright 2013.

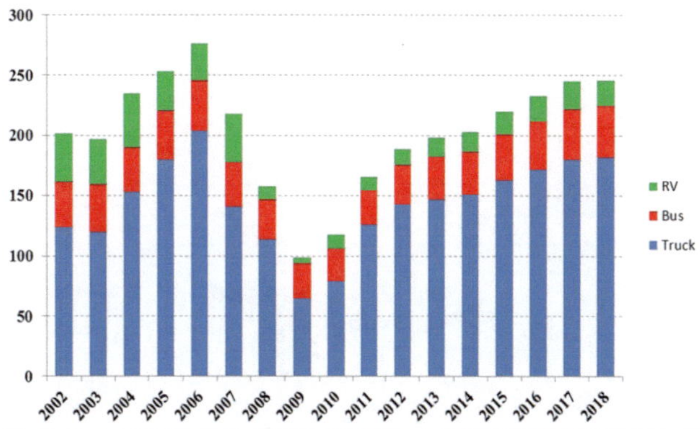

ANNEX FIGURE 4A-8 North American Class 5 thru Class 7 production. SOURCE: ACT Research.

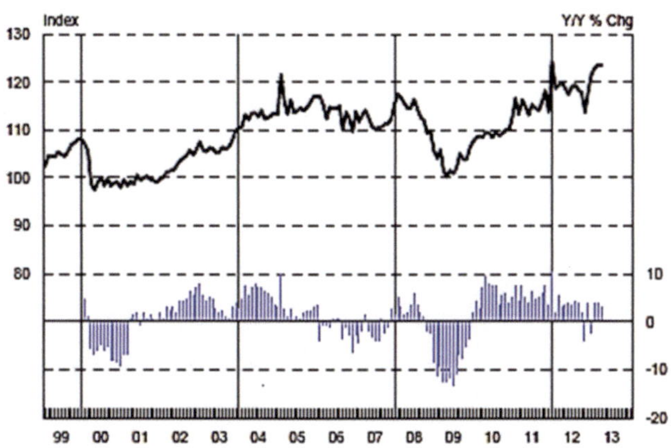

ANNEX FIGURE 4A-6 ATA truck tonnage index (seasonally adjusted). SOURCE: ACT Research, copyright 2013.

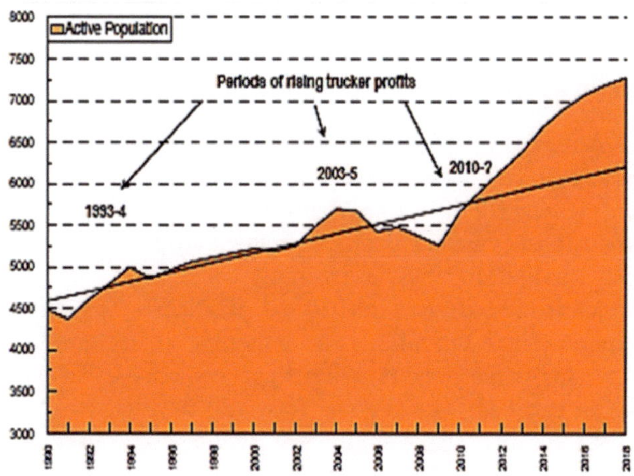

ANNEX FIGURE 4A-9 Class 8 population (freight per unit). SOURCE: ACT Research.

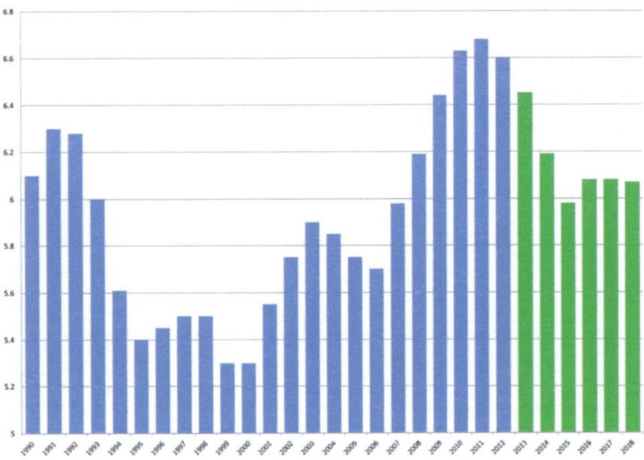

ANNEX FIGURE 4A-10 Average age of active population of U.S. Class 8 vehicles. SOURCE: ACT Research.

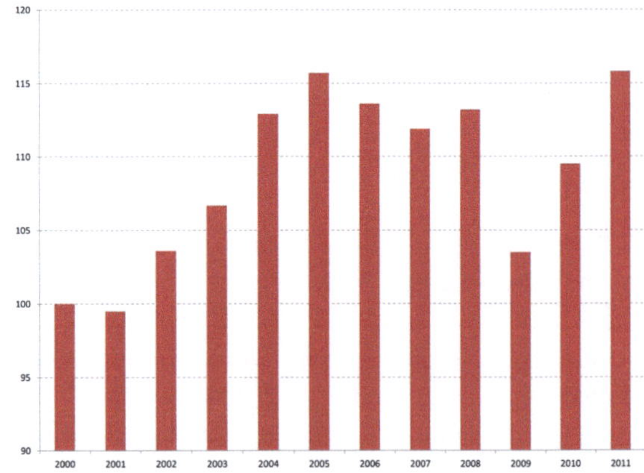

ANNEX FIGURE 4A-12 Tonnage index of ATAs for hire (seasonally adjusted). NOTE: Year 2000 = 100. SOURCE: ATA 2012b.

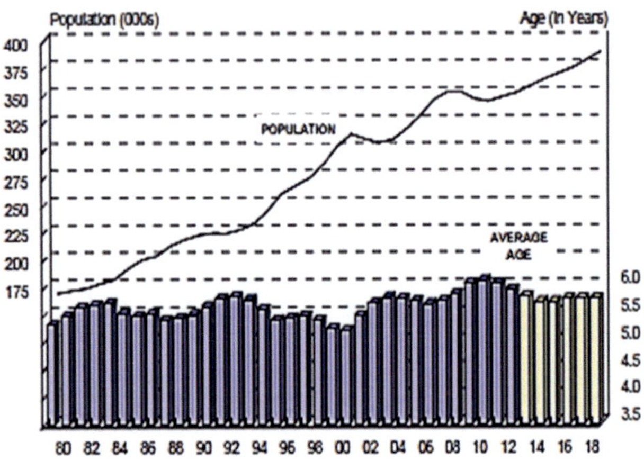

ANNEX FIGURE 4A-11 Population and average age of refrigerated vans. SOURCE: ACT Research.

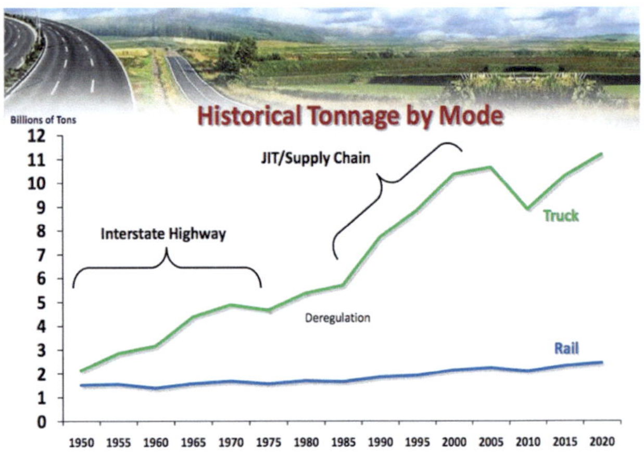

ANNEX FIGURE 4A-13 Historical tonnage by mode. SOURCE: ATA 2012a.

ATA Reports

American Trucking Associations is the premier trade association for fleets in the United States. Thousands of fleets are members of the various state associations. ATA compiles and publishes annual reports on freight transportation trends and activities. It publishes a monthly Truck Tonnage Report, which is used as input to FTR and ACT Research monthly reports. Since ton-miles is an important characteristic to track for existing and future fuel use regulations, ATA may be an excellent source going forward (see Figures 4A-12 to 4A-14 and Table 4A-3). From Figure 4A-13, used in a public webinar in 2013, it can be seen that tonnage has been tracked for many years. However, this information is currently aggregated and does not include the necessary detail by class of vehicle.

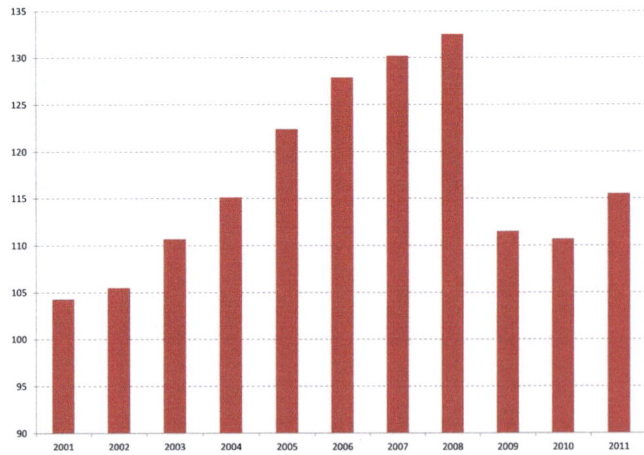

ANNEX FIGURE 4A-14 Index of less-than-truckload revenue per ton (seasonally adjusted) of ATAs for hire. NOTE: Year 2000 = 100. SOURCE: ATA 2012b.

ANNEX TABLE 4A-3 U.S. Retail Truck Sales (thousands)

	Class								
Year	I	II	III	IV	V	VI	VII	VIII	Total
2000	5,124.2	2,421.5	116.3	47.4	29.1	51.2	122.6	211.5	8,123.8
2001	5,112.8	2,507.7	101.5	52	24.4	42.4	91.6	139.6	8,072.0
2002	5,007.3	2,564.7	80	37.8	24	45.1	69.3	146	7,974.4
2003	6,268.5	2,669.2	90.8	39.6	29	51	66.8	142	9,357.0
2004	6,457.7	2,796.0	107.3	47.1	36.3	69.8	75.3	203.2	9,792.6
2005	6,585.8	2,525.0	166.9	48.5	46.3	60.2	88.9	252.8	9,777.3
2006	6,143.3	2,430.5	149.8	50.3	49.5	70	90.8	284	9,268.2
2007	5,682.3	2,662.6	165.9	51	44.9	53.8	70.4	151	8,841.9
2008	4,358.3	1,888.2	134.8	36.4	40.3	39.4	48.9	133.5	6,679.8
2009	3,527.9	1,305.8	111.7	19.9	23.9	22	39.1	94.8	5,145.1
2010	4,244.6	1,513.8	161.4	12.1	31	29.1	38.4	107.2	6,136.8
2011	4,714.1	1,735.6	195.3	10.5	40.7	41.2	41.2	171.4	6,951.2

SOURCE: ATA (2012b).

ATRI

As can be seen from a number of the preceding charts and graphs above, the American Transportation Research Institute (ATRI) is the trucking industry's not-for-profit research organization and is an autonomous member of the ATA Federation.[7] ATRI is a recognized source of data on the industry. The list of publications and areas of research are too numerous to list.

Wards Auto

Wards Auto is often cited as the source of information on production and sales, by vehicle size and brand. Often they have used R.L. Polk registration data for this information.[8]

R.L. Polk

R.L. Polk registration data are the recognized standard for information on commercial vehicles and their use.[9] This basic database is used by many organizations in trucking, by vehicle OEMs, and by suppliers to better understand volumes of trucks by class and vocation.

North American Council for Freight Efficiency

NACFE conducts an annual Fleet Fuel Study, where fleets provide a rolling 10 year history for the adoption of various technologies (see Annex Figure 4A-15). These data, although limited to a few adopters, could provide a solid, repeatable process for determining technology adoption rates. Annex Figure 4A-15 shows adoption rates over the years for various categories of technologies of all participating fleets.

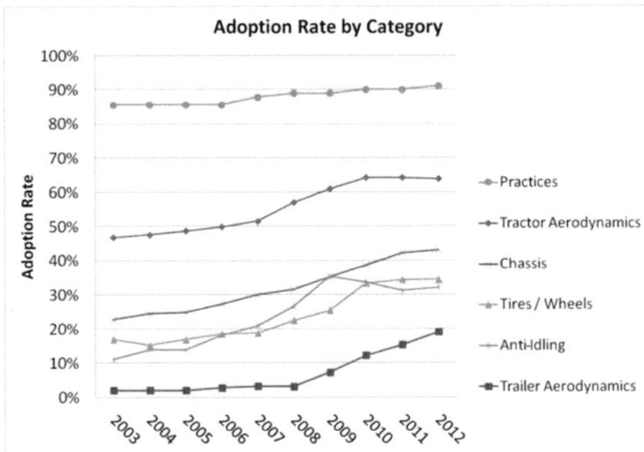

ANNEX FIGURE 4A-15 Rates at which various technologies were adopted. SOURCE: North American Council for Freight Efficiency. Available at http://nacfe.org/wp-content/uploads/2014/01/NACFE-2013-Study-Report-FINAL-March.pdf. Accessed December 11, 2013.

[7] The American Transportation Research Institute website is available at http://atri-online.org.

[8] The Wards Auto website is available at http://wardsauto.com.

[9] The R.L. Polk website is available at https://www.polk.com.

ANNEX 4B
ADDITIONAL WAYS TO OBTAIN INFORMATION IN THE FUTURE

Since the fuel economy and GHG regulations for MHDVs are new, it is important to establish methods and processes that will serve both industry and government in the coming decades. Some sources of information already exist, such as information on bills of lading, fuel purchases, and road use taxes. All such sources of information should be investigated for possible use on an ongoing basis for tracking items of interest to these regulations and their cost and their effectiveness. New sources of information are coming online. Federal and state governments have regulations requiring the reporting of numerous items that can be used for creating a baseline and regularly tracking the improvement in the industry in fuel usage and ton-miles. This includes hours of service, inspection reports, bills of lading, weigh station inspections, and fuel tax reporting. In the near future, the Federal Motor Carrier Safety Administration is expected to mandate the use of recorders for drivers' hours of service. These could become excellent sources for tracking information on the vehicles. California has a rule for electronic trailer refrigeration units that includes a telematics monitor requirement. Also, as mentioned in Chapter 2, the EPA SmartWay program is another source for fuel consumption information from participating fleets. As NHTSA deliberates on potential rules and regulations for V2X (Vehicle to Vehicle, Vehicle to Infrastructure), the regulatory instrument could be crafted to provide new sources of information on fuel use and GHG emissions. EPA and NHTSA should establish a plan for obtaining information in the future that takes account of all public, private, and governmental sources of information.

5

Natural Gas Vehicles: Impacts and Regulatory Framework

Natural gas (NG) accounts for about 25 percent of all energy use in the United States yet only 0.1 percent is used in transportation, equivalent to about 0.5 billion gallons per year of petroleum fuel. However, in the short time since the release of the report *Technologies and Approaches to Reducing the Fuel Consumption of Medium- and Heavy-Duty Vehicles* (NRC, 2010), NG has emerged as a viable option for commercial vehicles, driven by the rapid drop in its cost as a result of new technology. Yet, uncertainties exist relative to the extent and quality of reserves, safety and community acceptance, fuel cost and continued availability, sustainability, etc. This chapter starts with a brief review of why NG has emerged as a potentially significant fuel for medium- and heavy-duty vehicles (MHDVs). It then discusses the technology and infrastructure that will be needed for NG-fueled trucks, which can operate on either compressed natural gas (CNG) or liquefied natural gas (LNG). It describes the regulatory framework for greenhouse gas (GHG) emissions (primarily carbon dioxide, or CO_2) and fuel economy standards. Finally it presents the key findings and recommendations relevant to NG-fueled trucks.

SUMMARY OF SUPPLY AND DEMAND TRENDS FOR NATURAL GAS FUEL

To better understand what this potential means for the United States and its energy future, the National Research Council's (NRC's) Board on Energy and Environmental Systems (BEES) convened a meeting on September 11, 2012, assembling technical experts from private industry as well as representatives of government and academia. The primary goals were to assess the current state of NG development and to identify key research and technological gaps. This effort was augmented by presentations to a meeting of the committee on July 31 and August 1, 2013. This discussion is largely based on those forums.

NG production, driven by shale gas (Figure 5-1), began accelerating in 2005 and is expected to continue to grow (Figure 5-2). There are three key enabling technologies for the current boom: horizontal drilling, which was largely proven in the 1990s; geosteering, the use of real-time, on-site geologic data to guide the well bore; and hydraulic fracturing, or fracking, which involves the use of fracturing fluid (predominately water-based) to fracture the rock and proppants (e.g., sand) to hold the fractures open, allowing the trapped gas to flow through the shale and into the well.[1] These technologies have allowed the extraction of NG from huge reserves that previously had been economically inaccessible. The Interstate Natural Gas Association of America expects NG consumption to grow 1.6 percent per year.[2]

Increased production has driven prices down, which has led to increased demand, in particular for electricity generation and transportation fleet use. It has also benefited residential, commercial, and industrial customers. Manufacturing appears to be on the rebound in this country in part because of the low price of NG, which is used as both a fuel and a feedstock for petrochemicals. Potential NG production levels could exceed growing demand, and the United States could become a net gas exporter by 2015.[3] However, low current prices do little to encourage more drilling to expand production, and higher prices will dampen demand. It is not clear where supply and demand will balance.

With the recent, rapid increase in shale gas development has come increasing concern about its environmental impacts, both positive and negative. The substitution of NG for coal for electricity generation has been an important factor in reducing U.S. emissions of GHG. Horizontal drill-

[1] Stephen A. Holditch, Texas A&M University, "Shale gas development," Presentation to the Board on Energy and Environmental Systems, September 11, 2012.

[2] Donald Santa, International Natural Gas Association of America, "Pipelines make it possible," Presentation to the Board on Energy and Environmental Systems, September 11, 2012.

[3] Majida Mourad, Cheniere Energy, Inc. "LNG exports— How much and how soon," Presentation to the Board on Energy and Environmental Systems, September 11, 2012.

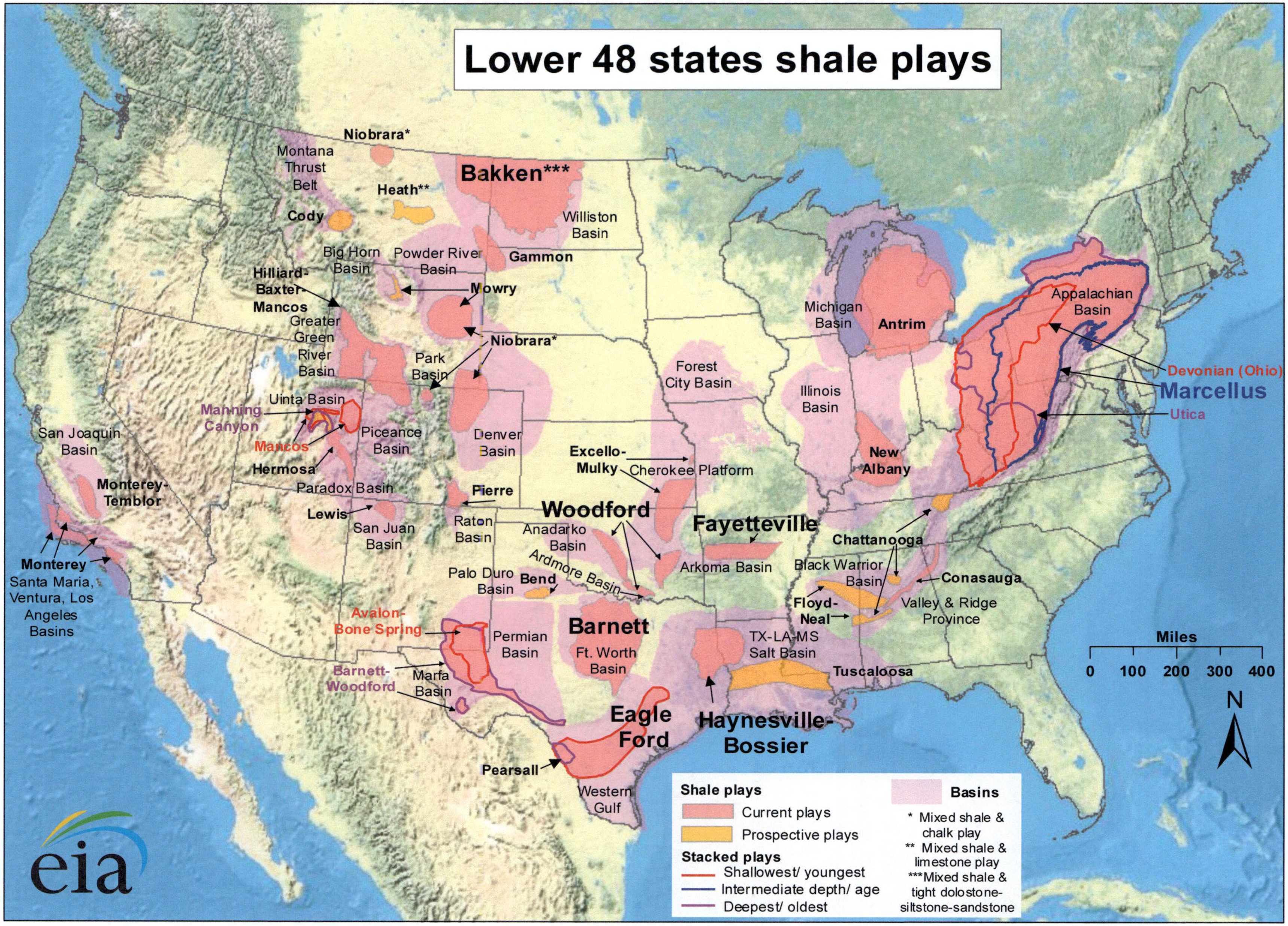

FIGURE 5-1 Shale gas plays in the continental United States. SOURCE: Energy Information Administration (EIA). Available at http://www.eia.gov/oil_gas/rpd/shale_gas.pdf. Accessed December 23, 2013.

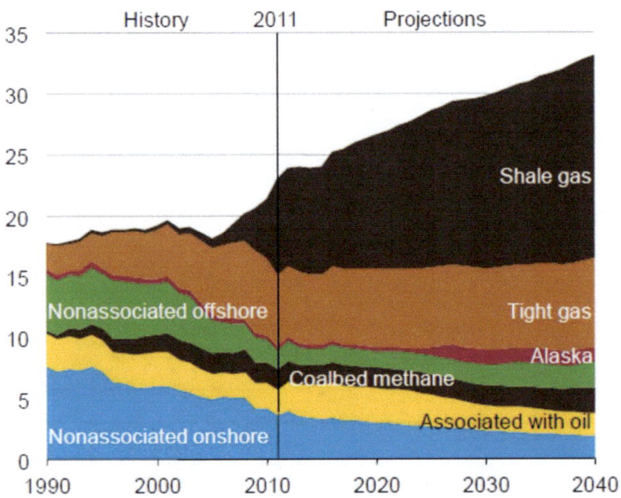

FIGURE 5-2 U.S. natural gas production. SOURCE: EIA (2013a).

ing causes much less surface disruption than conventional drilling—for example, Apache Corporation's Horn River Development uses a drill pad of just 6.3 acres to recover gas from approximately 5,000 acres. However, this means that local pollution (e.g., particulate matter, volatile organic compounds, and nitrogen oxides [NO_x]) may be more concentrated. In addition, NG largely consists of methane, which is a powerful GHG. Leakage, most of which is estimated to come from gas production activities, could negate the hoped-for climate benefits of reducing CO_2 emissions by replacing other fossil fuels with NG. Methane has a shorter lifetime in the atmosphere than carbon dioxide, but its higher radiative forcing—that is, its ability to redirect heat that would otherwise escape the atmosphere—means that over 100 years it has 20 times the GHG impact of CO_2. One analysis concluded that after taking into account current estimates of leakage, converting heavy-duty diesel trucks would have a net negative effect on climate change for centuries.[4] One estimate of gas leakage, based on measurements at 190 onshore gas production sites, is 0.42 percent of the total gas production (Allen et al., 2013). Note that this leakage exceeds the amount of NG currently used in transportation. Other estimates of fugitive emissions have been significantly higher (e.g., Howarth et al., 2011).

Water contamination is a widely discussed concern about hydraulic fracturing. This can be divided into two main issues: (1) underground contamination related to well integrity and (2) disposal of wastewater from the hydraulic fracturing process.[5] At the surface, an integrated management plan is needed to address the supply, handling, reuse, and disposal of the fracking fluid to ensure sustainability throughout the production cycle.

In the electric power sector, the low price of NG has directly caused the closure of coal plants, as it has become more economical to use combined-cycle NG plants (with thermal efficiencies up to 65 percent) for electricity production. However, fuel price is the dominant contributor to the cost of electricity (55 percent). One analysis concluded that the break-even fuel price is between \$4 and \$6 per million British thermal units (mmBTUs).[6]

In the heavy-duty transportation sector, price has a less direct effect on the use of NG as a fuel because delivering and compressing (or liquefying) the fuel account for a large share of the price at the pump. The break-even price of NG relative to diesel fuel is around \$6 per million BTU (predelivery, not at the pump). If the costs of NG vehicles themselves come down relative to the costs of their diesel counterparts (discussed in the next section), the break-even value could be as high as \$9 to \$12 per million BTU. If, as projected by the Energy Information Administration (EIA), the price of NG in 2035 is about \$7 per million BTU (EIA, 2013a), its use in the transportation sector will likely depend in part on future technological improvements.

Currently, the biggest obstacles to NG use for freight transportation are (1) the lack of widespread and dependable infrastructure, (2) the substantial increase in weight and cost of the fuel tanks compared to diesel tanks, and (3) the availability of NG vehicles, although almost all MHDV manufacturers now offer a NG engine. More detailed discussion of infrastructure and technology follows in a later section of this chapter. Pipeline and infrastructure investment in the United States and Canada is likely to exceed \$200 billion over the next 25 years (see footnote 2).

EIA expects increased production, lower imports, higher exports, and higher prices, as shown in Table 5-1.

NATURAL GAS ENGINES AND VEHICLES

Technology

Overview

NG internal combustion engines are a well-developed and established technology. There are over 11 million NG vehicles worldwide, including passenger vehicles. In the United States, NG-fueled MHDVs, especially transit buses,

[4] Steven Hamburg, Environmental Defense Fund, "Methane leakage from natural gas production, transport and use—Implications for the climate," Presentation to the Board on Energy and Environmental Systems, September 11, 2012.

[5] Mark Boling, Southwestern Energy, "Forum on Unconventional Natural Gas Issues: Water Quality," Presentation to the Board on Energy and Environmental Systems, September 11, 2012.

[6] Revis James, Electric Power Research Institute, "The Role of Natural Gas in the Electricity Sector," Presentation to the Board on Energy and Environmental Systems, September 11, 2012.

TABLE 5-1 Projections of Domestic Production and Prices for Natural Gas and Diesel Fuel Price

	Year		
	2012	2025	2040
Production (quadrillion BTU)	24.59	32.57	38.37
Henry hub price ($)	2.75	5.23	7.65
Price delivered to transportation ($)	14.64	15.57	19.67
Diesel fuel price ($)	8.80	29.02	34.53

NOTE: Prices are per million BTU in 2012 dollars. Note that a million BTU is equivalent to about 8 gallons of diesel fuel. Thus, natural gas costs on the order of $2.00 per gallon equivalent, much less than diesel fuel.
SOURCE: EIA (2013a).

have been incentivized for roughly 20 years in some states as part of emission-reduction programs. For MHDVs to use natural gas fuel, the most significant differences from current vehicles are the onboard fuel storage method and, for compression ignition (diesel-fueled) vehicles, the means of introducing and igniting the fuel in the engine. On-vehicle storage is either by high-pressure (3,600 psi is typical) CNG cylinders or by cryogenic containers filled with LNG. An illustration comparing on-vehicle storage of NG with diesel is shown in Figure 5-3.

For the same truck mission, the CNG tank plus fuel weighs about four times as much as a diesel tank plus fuel. LNG tanks and fuel weigh about twice as much as diesel. The cost of either CNG or LNG storage adds $40,000 to $50,000 to the cost of a heavy truck, but with the current low price of NG, the payback period for long-haul trucks is on the order of only 2 years.

There are three general technical classifications of NG engines, as shown in Table 5-2. Either CNG or LNG can replace gasoline with only modest changes to the spark ignition (SI) engine. Compression ignition (CI) engines are more complicated; NG can be used in combination with diesel fuel (dual-fuel); or it can supply all the energy to a high-pressure direct-injection (HPDI) CI engine, in which a small amount of diesel fuel is needed to achieve ignition.

Table 5-2 notes several advantages and disadvantages of each configuration.

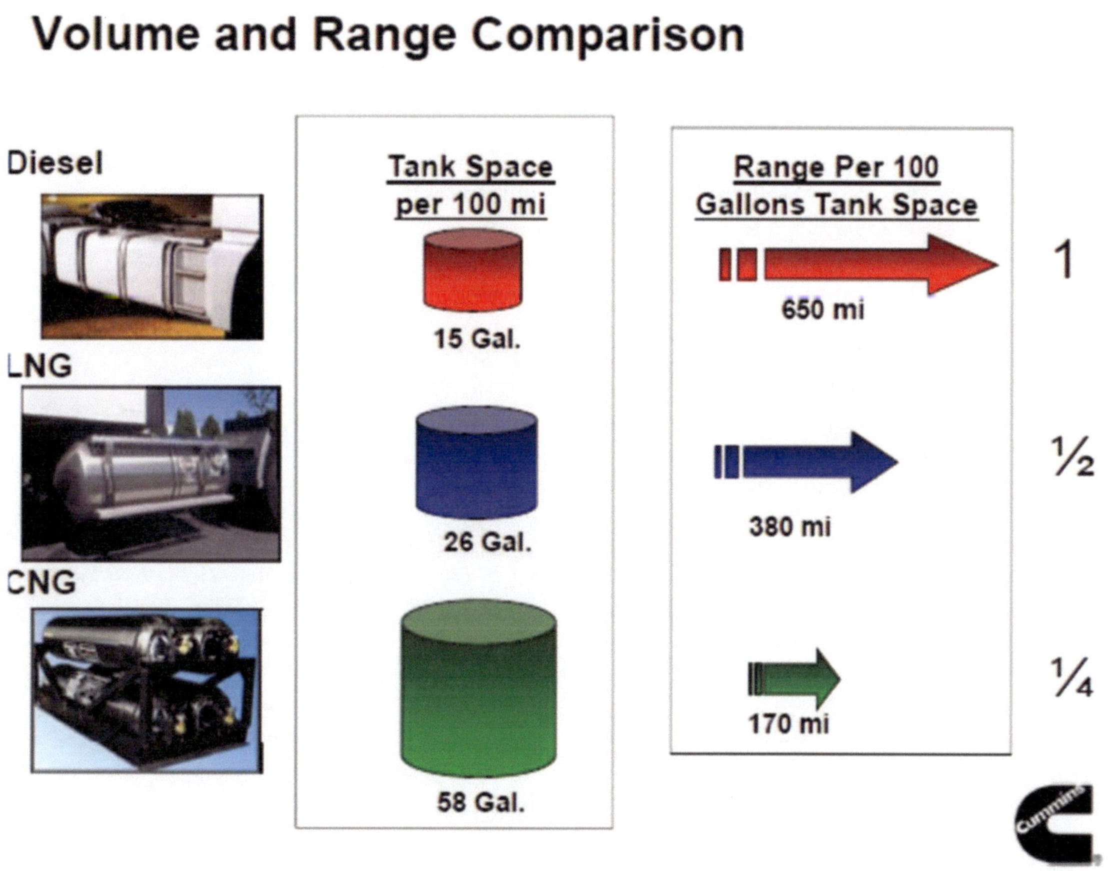

FIGURE 5-3 Comparison of diesel and NG fuel tanks. SOURCE: John Wall, Cummins, Inc., "Opportunities and barriers for natural gas: Expanding natural gas in transportation," Presentation to the Board on Energy and Environmental Systems on September 11, 2012.

TABLE 5-2 Natural Gas Engine Technology Comparison

	Dedicated NG	Dual Fuel	HPDI
Technology	Spark ignition Throttle to control A/F Lean or stoichiometric/EGR Two-/three-way catalyst A/T	Compression ignition With or without EGR Diesel + premixed NG DOC DPF+SCR A/T system	Compression ignition EGR Diesel pilot + in-cylinder NG injection Not dual fuel—all power from NG DOC DPF + SCR A/T system
Advantages	100% diesel substitution Lowest emissions v. other NG technologies Simple, passive aftertreatment LNG or CNG	Power density similar to diesel Efficiency similar to diesel No spark plugs Potentially retrofittable on existing diesel engine Can run on diesel only	Power density similar to diesel Efficiency similar to diesel No spark plugs 95% diesel substitution
Disadvantages	Spark plug life 10-15% lower efficiency than diesel, better efficiency than gasoline PFI Lower power density Knock limited	Substitution limited to 50-80% Unburned methane emissions Misfire at light loads Knock at high loads	High system cost/complexity Requires both fuels, always LNG + cryogenic in-tank pump for high-pressure injection

NOTE: A/F, air:fuel ratio; A/T, aftertreatment; DOC, diesel oxidation catalyst; EGR, exhaust gas recirculation; PFI, port fuel injection; SCR, selective catalytic reduction; DPF, diesel particulate filter.
SOURCE: Tim Frazier, Cummins Westport, "Cummins Westport Natural Gas-Fueled Engines," Presentation to the committee on July 31, 2013.

These configurations of SI, pilot ignition, and dual fuel have existed for many years, but under very different emissions standards. At least 11 different engine displacements for CNG/LNG MHDVs are expected to be available by 2015. As more original equipment manufacturers (OEMs) are introducing NG options to their product lines, the share of CNG/LNG MHDVs continues to grow.

Details of seven engines are presented in Table 5-3. They range in displacement from 6.0 L to 12.7 L. The engines are predominantly SI, capturing the simplicity of fuel/air mixing with stoichiometric combustion and facilitating criteria emission control with the similarly simple three-way catalyst (TWC). These NG engines typically are based on diesel engines (not SI) to achieve the durability and reliability expected for trucking applications. For example the Cummins-Westport engine has over 80 percent of its parts in common with the base diesel-fueled engine (ISX12). However, the SI combustion system is less efficient than CI engines by roughly 15 percent. One manufacturer expects to offer a HPDI compression ignition model that should approach diesel engine efficiency. A Westport Innovations 15L-HPDI sold for about 7 years but was taken off the market in October 2013; it does not appear in the table.

Figure 5-4 depicts the types of NG-fueled vehicles that are available and the examples of engines for those vehicle classes. The engines shown from Ford and GM are conversions of OEM-built vehicles where, upon delivery, customers use an OEM-qualified vehicle modifier that supplies the fuel tanks, lines, and special fuel injectors. Depending on the size of the tank, these modifications can cost from $6,000 to $9,500. The NG-fueled vehicles offered by Ford, GM, and Chrysler are all derived from gasoline SI engines and can operate as bi-fueled configurations. Bi-fuel indicates that vehicles are equipped with both CNG and gasoline fuel systems; switching from gas to gasoline is automatic. This process can significantly extend vehicle range when CNG refueling is not available. Certain of the heavier vehicles are equipped only with CNG fuel systems. Figure 5-4 does not include aftermarket conversions procured after the sale of the vehicle.

As of 2010, the number of medium- and heavy-duty NG vehicles in the United States is estimated to have been between 30,000 and 50,000, out of roughly 10 million total MHDVs (TIAX for American Natural Gas Association [ANGA]).

Fuel Consumption and GHG Comparisons of Natural Gas and Diesel Engines

When a fuel is combusted, its CO_2 release per unit heat released is a function of its carbon content. Because it has a relatively low carbon content, NG releases about 28 percent less CO_2 per BTU of heat than diesel fuel. However, SI engine efficiency is considerably lower than that of mature diesel engines, especially at light loads, partially offsetting the inherent GHG benefit of NG. In addition, unburned methane may be emitted by the vehicle, and upstream emissions from the production and delivery of the NG must be considered in a well-to-wheels comparison. Methane is a very potent GHG, so if these emissions are significant, NG vehicles could contribute more to GHG emissions than diesel vehicles.

Several studies have compared CO_2 emissions of diesel and NG vehicles and engines in driving and test cycles.

TABLE 5-3 Features of Available Natural Gas Engines

Engine Description / Parameters	Vortec 6.0L V8 gasoline conversion bi-fuel[1]	ISB6.7-G purpose built diesel derivative	PSI 8.8L-G purpose built	ISL-G purpose built diesel derivative	Doosan GL11 purpose built	ISX12-G purpose built diesel derivative	D13-LNG diesel derivative diesel pilot
Manufacturer	GMC	Cummins	Pow er Solutions	Cummins	Doosan	Cummins	Volvo
Vehicle Vocational Target	CL 2b-6 Van, Trk	CL6 Trk,SchlBus	CL 6-8 Truck	CL6-8 Truck,	CL6-8 Truck,	CL 8 Truck, Bus	CL 8 Truck
Expected NG form	CNG	CNG or LNG	CNG	CNG or LNG	CNG	CNG or LNG	LNG
Number ofCylinders	V8	6	V8	6	6	6	6
Displacement (L)	6.0	6.7	8.8	8.9	11.1	11.9	12.7
Displ/Cyl (L)	0.75	1.1	1.1	1.5	1.8	2	2.1
Combustion System	SI-stoic	SI-stoic	SI	SI-stoic	SI-stoic	SI-stoic	CI
Compression Ratio					>10.5	12.1	15.6
Fuel System Description	Port injection	Electronic throttle body mixing	Electronic throttle body mixing	Electronic throttle body mixing	Electronic throttle body injection	Electronic throttle body mixing	LNG-diesel dual fuel injectors; diesel pilot
Air Handling System Description	naturally aspirated	WG-Turbo; CAC	Turbocharged	WG-Turbo; CAC	Turbocharged, CAC	WG-Turbo; CAC	VGT; CAC
EGR (Y/N)	N	Y-HPL	N	Y-HPL	Y-LPL	Y-HPL	Y
Aftertreatment Descrptn: 2010 Criteria Emissions	TWC	TWC		TWC	TWC	TWC	DOC,DPF & SCR
Rated RPM	5000	2600	3400	2200	2200	2100	1700
Rated kW/L	28.1	36.1	22.7	26.9	22.9	25	27.9
Rated kW(HP)	225'(301)	242'(325)	200'(268)	239'(320)	253'(340)	298'(400)	354'(475)
Max Torque N-m(lb-ft)	452'(333)	1017'(750)	1356'(1000)	1356'(1000)	1627'(1200)	1966'(1450)	2373'(1750)
Issues & Unique Barriers	bi-fuel						both fuels, alw ays
Production available	Y	2015	mid 2014	Y	2014	Y	mid 2014
Resource	I	E	F	B	G	B	D

Notes: [1] bi-fuel means non-simultaneous fuel (natural gas and gasoline) supply to engine

While the results differ somewhat, NG engines and vehicles generally emit about 5 to 20 percent less CO_2 (Krupnick, 2010; Kamel et al., 2002; Greszler, 2011). The advantage for NG is very dependent on the drive cycle. One estimate of the impact of methane emissions, shown in Figure 5-5, is that they reduce the CO_2-equivalent GHG emissions benefit of NG from 13 percent to only 5 percent.

This review of the data comparing diesel and NG engines affirms that the chemical advantage of NG for low GHG emissions is largely (but not completely) offset by the lower efficiency of most NG-fueled heavy-duty engines. This will need to be considered in setting specific GHG and fuel consumption standards for medium- and heavy-duty vehicles using NG, as was done in setting different standards for gasoline and diesel engines.

Safety: On-Vehicle NG Storage

CNG has been used successfully for many years to power commercial vehicles operating in limited-range applications, usually in urban areas. With the recent increase in the availability of low-cost NG, it is anticipated that its use in long-haul Class 7 and Class 8 trucks will increase, especially using LNG as a means of extending vehicle range. Crash-related safety risks associated with NG storage in larger vehicles compared with diesel fuel risks do not appear to be appreciably different. However, because the number of NG-fueled commercial vehicles is small compared with diesel-fueled vehicles, that conclusion is uncertain at this time. Furthermore, because public crash databases do not contain information on fuel type, they cannot provide categorical insight on risk.

In tractor vehicles, diesel tanks are often mounted on the outside of the tractor frame rails beneath the driver and occupant access doors. This location is exposed to side impacts from other vehicles, side impacts with objects, truck rollovers, and opposite direction sideswipe crashes. In single unit trucks (SUTs), the tanks are often located below the bed, which is more protected.

If a NG tank is ruptured and the gas is ignited, the fire will be appreciably different from a fire involving diesel fuel. The risk to the driver and occupants from such events is not well understood due to a lack of data and documented crash experience.

LNG is at much lower pressure and would not ignite so violently, but the tank shells are more complex to provide thermal insulation and may be more vulnerable to puncture than high-pressure CNG tanks. The extreme cold (–260 °F) of LNG presents risks of its own. There is little information on accidents involving NG trucks because so few are on the road, but experience with diesel trucks provides some useful

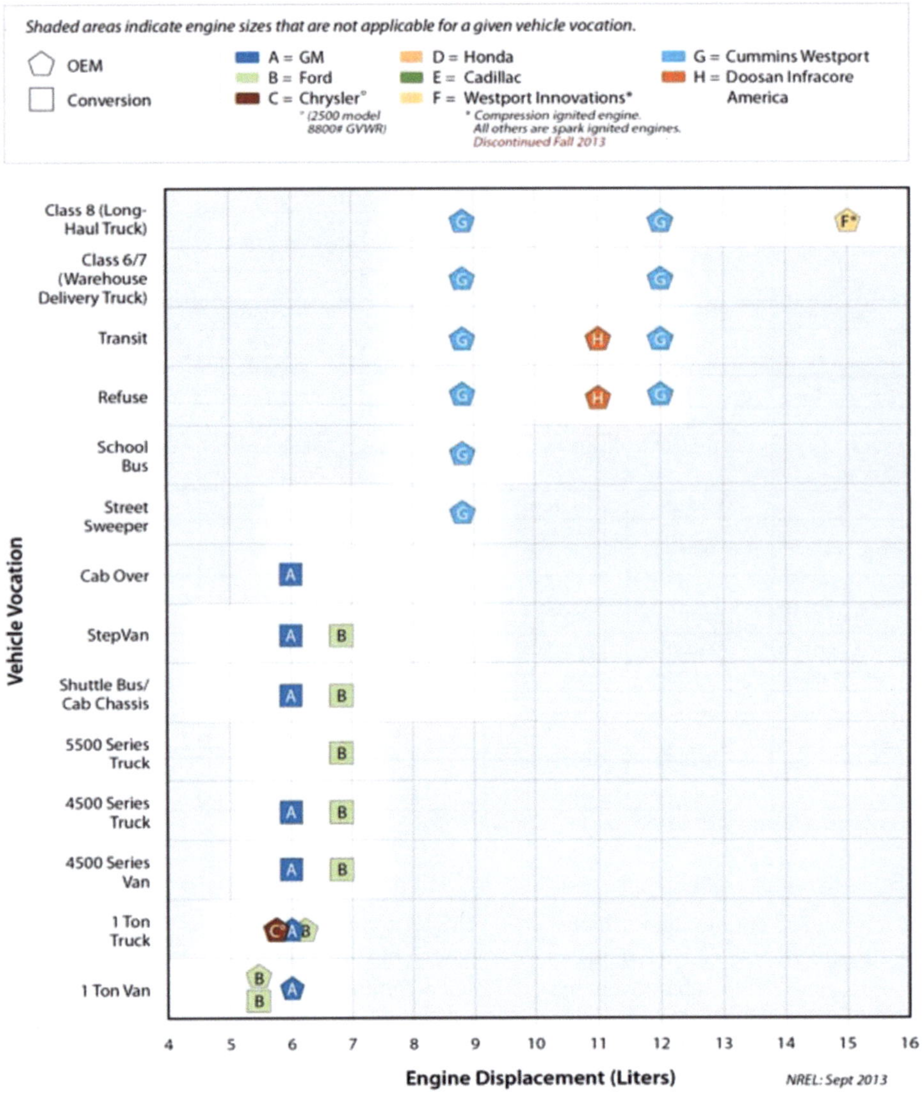

FIGURE 5-4 Available natural gas vehicles and OEM-authorized conversions. SOURCE: NREL.

information. In particular, fires (from all causes, not just fuel) are involved in accidents responsible for 14.2 percent of all fatalities and serious injuries in tractors, but only 4.8 percent in SUTs (UMTRI, 2013). These data suggest that protecting NG tanks in tractor applications will be very important for safety. Rollovers are the most frequent source of fatalities for both vehicle types, but that is likely to be true no matter which fuel is used. Heavy trucks are 30 to 40 times heavier than passenger vehicles. When crashes occur, they tend to be very damaging to cab structures and auxiliary components, particularly when large fixed objects are encountered.

There are various NG tank options available for large truck applications. Type 4 CNG cylinders have a "plastic core and are fully wrapped with a composite, such as carbon fiber. Other less expensive options are Type 2 and Type 3 CNG tanks, which have a steel or aluminum core and are composite wrapped."[7] For LNG-fueled vehicles, Dewar flask cryogenic tanks are used to keep the fuel in its liquid state. An LNG fuel tank can hold between 56 and 80 diesel-equivalent gallons. LNG is the preferred form of NG fuel if the daily operating range is over 400 miles. Given the diversity of NG tanks available for the heavy truck market, the crash characteristics, and associated risks, a review of technical standards and safety-related issues appears to be warranted, particularly with respect to tank type, location, and shielding. At present, there are two SAE-recommended practice

[7] Kenworth, "Kenworth Truck Company Offers Advice on Spec'ing for Natural Gas Power." Available at http://www.kenworth.com/news/news-releases/2011/december/kenworth-truck-company-offers-advice-on-specing-for-natural-gas-power.aspx. Accessed March 12, 2014.

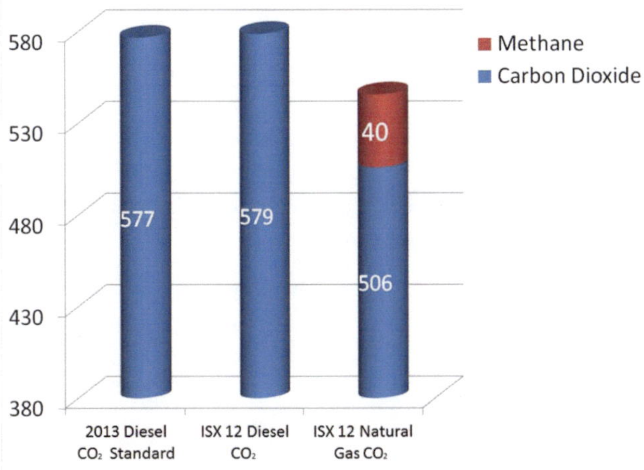

FIGURE 5-5 Comparison of emissions from natural gas and diesel engines. NOTE: bhp, brake horse power; g CO_2, grams carbon dioxide. Natural gas emissions were calculated as follows: CO_2 engine emissions + [(CH_4 engine emissions – CH_4 limit) * global warming potential] = 506 + [(1.7 – 0.1) * 25] = 546. The Federal Test Procedure (FTP) was employed. SOURCE: Tim Frazier, Cummins Westport, "Cummins Westport natural gas-fueled engines," Presentation to the committee on July 31, 2013.

standards for medium- and heavy-duty NG vehicles. SAE J2343 focuses on LNG while SAE J2406 addresses CNG. SAE J2343 refers to tank placement, but neither standard covers tank shielding requirements to prevent punctures during crashes. Industry could benefit from best practice directives to minimize safety risks associated with NG fuel tanks on commercial vehicles.

Technology Improvement Opportunities

Although the engines available today are robust and capable, improvements could further increase efficiency and reduce maintenance costs. Areas where further R&D for NG engines would be beneficial are listed in Table 5-4.

It has been suggested that in addition to needed advancements in engines and aftertreatment, a uniform national standard for NG quality for vehicular use is needed. Higher carbon content hydrocarbons in LNG tanks can lead to fluid stratification, and the composition of CNG tanks can affect autoignition, combustion behavior, and engine power density. The Engine Manufacturers Association recommends a minimum methane number of 80 for CNG, a maximum inert content of 4 percent, and a 16,200 BTU/lb minimum energy content, in addition to limits on contaminants such as water and sulfur compounds, among others (EMA, 2010). Activities to support standard setting are in progress in the Society of Automotive Engineers (SAE) and the Coordinating Research Council (CRC), and ISO-15403 has been published as a European standard for NG vehicle fuel (Kauling, 2013). Assurance of a minimum quality for in-use compressed and liquefied motor vehicle NG fuel could allow for the design of more efficient and less costly engines. However, a standard for in-use motor vehicle NG could require further processing of some pipeline gas before compression or liquefaction. Given the limited demand for motor vehicle NG currently envisioned, compared to nationwide consumption for other purposes such as home heating and power generation, it is not clear if suppliers of NG would invest in the gas cleanup needed for motor vehicle fuel. This could result in NG fueling stations being unavailable in certain areas where pipeline gas does not meet motor vehicle fuel specifications. This could impede the increased use of NG as a motor vehicle fuel.

Finally, as noted above, the fuel tank, whether for CNG or LNG, accounts for most of the cost increment for NG vehicles over equivalent gasoline or diesel vehicles. In addition to cost, weight can be an issue. The cheapest solid steel (Type 1) cylinders weigh four to five times as much as gasoline or diesel tanks of the same capacity; advanced (Type 3) cylinders with thin metal liners wrapped with composite weigh about half as much as Type 1 tanks, although they cost more. Tanks with polymer liners weigh even less, but are even more expensive. Higher pressure tanks (up to 10,000 psi) could reduce fuel storage space, but at added cost and increased energy required to compress the gas.

While tank costs are likely to come down with increased and improved production, large cost reductions will depend on advanced technology. In the future, it may be possible to store CNG at higher density, even at 500 psi (within the 200-1,500 psi range for gas in NG transmission pipelines) in adsorbed NG (ANG) tanks using spongelike materials such as activated carbon. If successful, this technology could allow vehicles to be refueled from the NG network without extra gas compression, reducing cost and energy use and allowing the fuel tanks to be lighter (adapted from NRC, 2013). Also, at lower pressure, the shape of the tank can be adjusted as needed to fit the space available, thus minimizing the impact on cargo space (PNNL, 2010).

TABLE 5-4 Technologies for Improvements in Heavy-Duty Natural Gas Engine Brake Thermal Efficiencies and GHG Emissions for the Phase II Rule

Base Engine	Aftertreatment
Improved air handling	Improved three-way catalyst
High-efficiency turbo	Better methane oxidation
Low pressure loss EGR	Increased oxygen storage
Low restriction ports	
	Improved controls
Higher energy ignition	Advanced air/fuel ratio dithering strategies
Improved combustion	Adaptive algorithms
New bowl geometry	Advanced sensors
Lower swirl/higher squish	
	Crank case ventilation
Improved friction and parasitics	

SOURCE: Tim Frazier, Cummins Westport, "Cummins Westport natural gas-fueled engines," Presentation to the committee on July 31, 2013.

Infrastructure

CNG Infrastructure

NG is moved throughout the United States in an extensive network of pressurized pipelines. According to the American Gas Association (AGA, undated), there are 1.9 million miles of CNG distribution lines and an additional 300,000 miles of transmission lines. The transmission lines are for long-distance interstate transport and operate at high pressure, from 200 to 1,500 psi. The distribution and service lines to homes operate at low pressure, approximately 50 psi to less than 1 psi. CNG refueling stations for vehicles are connected to points in the distribution pipeline network. The gas industry spends over $6 billion per year on the transmission lines and $4 billion per year expanding the distribution system.

There are 632 public CNG vehicle refueling locations in the United States (AFDC, undated) and around 1,200 including stations for private fleets (Weeks, 2013). The number of fueling stations actually peaked in 1997 and then declined until 2008. The rate of growth of CNG stations has been about 11 percent per year for the last few years (Clay, 2013), probably driven more by increased use in trucks than in passenger vehicles. The infrastructure hurdle for freight-hauling heavy-duty NG vehicles appears more manageable than for passenger vehicles and for Class 2b vehicles because far fewer stations will be needed for servicing trucks.

There are two basic types of CNG fueling systems: *fast-fill* and *time-fill,* shown in Figure 5-6 (AFDC, undated). Both are capable of filling the on-vehicle tank to approximately 3,600 psi, but as the name implies, the fast-fill can refuel a passenger vehicle in a few minutes by supplying gas from high-pressure storage vessels that are maintained full by compressors. The time-fill system in essence refuels the vehicle over a long period (hours) by the CNG compressor's relatively small direct output, with small storage tanks used only for buffering. The time-fill stations are generally less costly to set up and are likely to be most appropriate for fleets that can refill overnight. Some stations may have both systems.

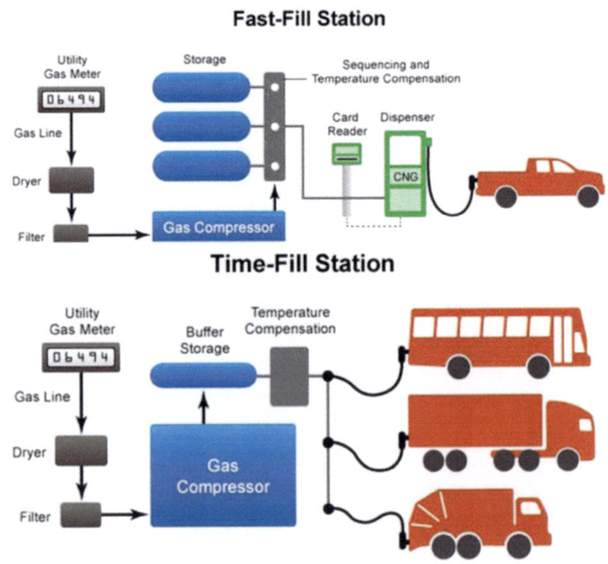

FIGURE 5-6 Conceptual diagram of compressed natural gas filling stations. SOURCE: http://www.afdc.energy.gov/fuels/natural_gas.html.

CNG stations are estimated to cost $600,000 to $1 million, with time-fills in the lower part of the range and fast-fills in the upper (ANGA, undated). These costs will be a considerable hindrance to the growth of CNG as a passenger vehicle fuel. One study estimated that between 16,000 and 32,000 stations would be needed to support a thriving NG vehicle population, with a much smaller number of refueling sites needed for heavy trucks (ANGA, undated). Equipping another 9,000 public stations will cost in the range of $5.4 to $9 billion.

LNG Infrastructure

LNG use is growing in transportation, especially for Class 8 freight trucks operating over interstate routes. LNG is formed by cooling NG to −162°C (−260°F), increasing its volumetric energy density about 600-fold and allowing it to be stored at relatively low pressure in highly insulated containers. The on-vehicle storage volume and weight advantages compared to CNG are considerable, but diesel fuel still has about 1.7 times the volumetric energy density of LNG. As shown in Figure 5-3, the approximate range per 100 gallons in a long-haul truck is: 650 miles (diesel); 380 miles (LNG); and 170 miles (CNG). LNG fueling stations are similar in configuration and operation to gasoline and diesel fuel retail outlets. LNG is delivered to the fueling station in tanker trucks, stored there, and dispensed into vehicles with cryogenic LNG storage tanks (see Figure 5-7). Many LNG fueling sites supply CNG as well.

The main disadvantage of LNG is that it gradually boils off as ambient heat penetrates the tank no matter how well insulated it is. This may not matter much in long-distance

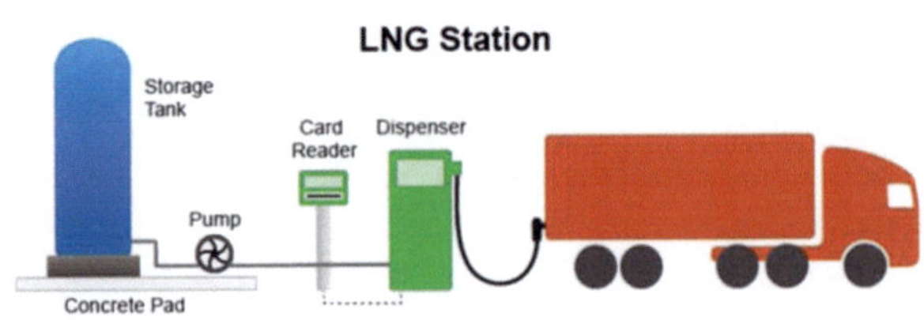

FIGURE 5-7 Conceptual diagram of LNG filling stations. SOURCE: http://www.afdc.energy.gov/fuels/natural_gas.html.

trucks that operate almost continually, but it probably precludes the use of LNG in light-duty vehicles.

Only about 40 nonprivate LNG fueling stations are in operation in the United States, many of them in California (AFDC, undated; TIAX, 2009). Clean Energy Fuels Corporation has been establishing a cross-country network of LNG fueling stations ("America's Natural Gas Highway"), with near-term plans for 100 LNG stations. Shell is working with TravelCenters of America to offer LNG for highway trucks and has a joint cooperation agreement on LNG with Volvo. LNG is produced at only about 50 to 60 sites in the United States, and there are a few LNG import terminals, so transport distance to LNG dispensing stations could be a detriment. California boasts the world's largest facility for producing LNG from landfill gas.

Conversion of NG to Other Fuels

NG can be converted to other fuels using well-known processes and technology. Each has advantages and disadvantages for storage, GHG impact, and cost:

- Dimethyl ether (DME)
- Methanol
- Ethanol
- Gas-to-liquids (GTL)
- Ammonia
- Electricity
- Hydrogen (for fuel cell vehicles)

DME-fueled truck development and demonstration is being carried out by Volvo with the fuel developer Oberon, which has a technology for converting NG or other feedstocks to DME via methanol as an intermediate. DME has storage characteristics similar to propane and much lower storage pressure and higher energy density than CNG. In a CI engine its performance is similar to that of diesel fuel.

Most methanol produced in the United States is derived from NG. In the mid-1980s, heavy-duty bus engines that ran on pure methanol or M85 (a mixture of 85 percent methanol and 15 per cent gasoline) were produced and emissions-certified, but subsequently CNG buses largely replaced them (Aldrich, 1995). Passenger vehicles were also produced and emissions certified to run on M85. A substantial infrastructure for dispensing M85 was established for fueling flex-fuel passenger vehicles, especially in California, but the program came to an end around 2005. Today in China, methanol is blended at low percentages in gasoline.

Although at present motor fuel ethanol comes from fermentation of biomass, it can also be produced from petroleum and NG. Recently several companies announced the development of new processes for the production of ethanol from NG. For example, Celanese, one of the world's largest producers of acetic acid from fossil fuels, is building industrial-scale ethanol plants in China and Indonesia using coal as a feedstock. The company also has a demonstration-scale ethanol plant in Houston, Texas, that uses NG as a feedstock, although Celanese (2013) has noted obstacles to supplying its ethanol for either E85 or E10 in the U.S. market. Coskata, Inc., is also developing a process for producing ethanol from NG. The proprietary Coskata process begins with the reforming of methane to produce synthesis gas, followed by fermentation of the synthesis gas to produce ethanol. Both Celanese and Coskata[8] estimate that they can produce ethanol at a substantial discount to gasoline or diesel fuel on an energy-equivalent basis

Converting NG to liquid hydrocarbons is a well-established technology relying largely on the Fischer-Tropsch process developed in the 1920s-1940s. Liquid fuels, especially diesel, from GTL technologies have been shown to be compatible with existing vehicles and can be combined with petroleum products, averting the need for a completely new dispensing infrastructure and a new vehicle technology to allow the widespread use of NG for transportation. GTL plants have been established worldwide where low-cost NG is available. To be profitable, the scale of GTL plants is enormous as is the capital investment, and the production of high-value chemicals in addition to fuel is important. Shell and Sasol are the largest GTL producers. Operating since 2011, Shell's Pearl GTL facility in Qatar is one of the largest such plants in the world (140,000 BPD products), nearly 10 times the size of the demonstration facility Shell built in Malaysia in the early 1990s. Products from the Pearl facility are shipped to the United States. Low-cost NG in the United States gives rise to consideration of domestic GTL production. Companies such as Compact GTL and Velocys are developing relatively small, modular GTL production technology to support operations of, say, less than 5,000 BPD. A 1,000 BPD commercial production compact GTL plant is being considered for construction in Pennsylvania, in the heart of one of the large shale gas plays.

NG is the source for 95 percent of hydrogen production in the United States, and fuel cells are candidates for certain heavy vehicles such as buses and drayage tractors. In California, fuel cell heavy vehicles are a key option where zero-emission vehicles are needed. There are 10 hydrogen refueling stations in the United States, most of them in California (AFDC, undated).

NG is a feedstock for anhydrous ammonia, a feasible engine fuel that can be stored as a liquid at pressures similar to propane. Although it has been demonstrated in both SI and CI combustion systems, its toxicity and acute incompatibility with the human body make its widespread use as a transportation fuel impractical.

About 30 percent of U.S. electricity comes from burning NG (EIA, 2013b), and this fraction is growing rapidly. Combined-cycle (gas turbine/steam turbine) technology can

[8] Coskata, Inc. "Natural Gas." http://www.coskata.com/process/index.asp?source=984A0E13-B27D-4267-8411-AA86E0CAA39F.

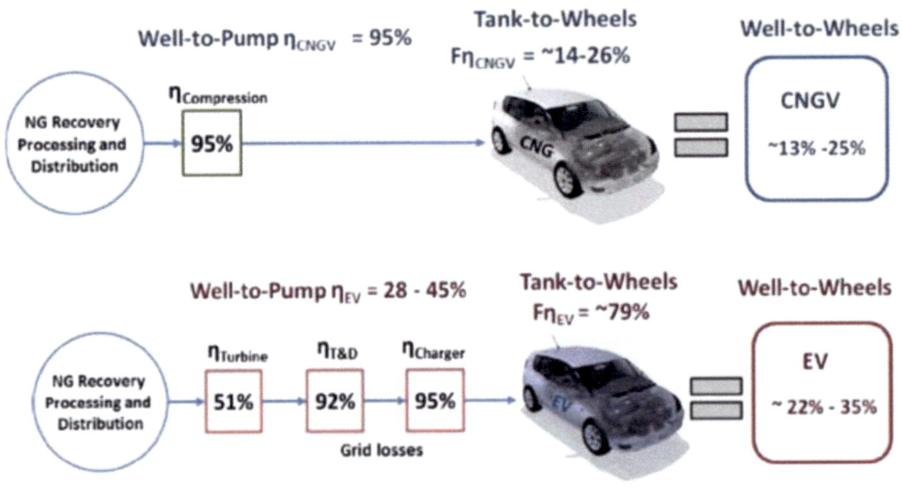

FIGURE 5-8 Comparison of efficiencies of compressed natural gas vehicles (CNGVs) and battery electric vehicles (BEVs). SOURCE: Curran, 2013.

be highly efficient. Some units are over 60 percent efficient, much higher than coal-fired generation. Electric vehicles (battery electric or plug-in hybrids) can take advantage of NG in this path, thereby displacing petroleum.

The conversion processes for all of the alternative fuel options described above emit GHGs. Therefore the processes that produce hydrocarbon fuels will result in higher total GHG emissions than NG-fueled engines, which are on par with or slightly better than comparable diesel-fueled engines. Only electricity and hydrogen from NG can produce lower GHG emissions than direct combustion of NG in engines due to the inherently high efficiency of battery electric and fuel cell vehicles.

Figure 5-8 compares the efficiencies from well to wheels of CNGVs and electric vehicles (the figure was developed for automobiles, but the well-to-pump numbers would be the same for trucks). While today's CNGVs might attain a total well-to-wheels efficiency of 25 percent, the battery electric vehicle could attain 35 percent, which would result in about one-third less GHG emissions.[9]

This figure assumes efficient (51 percent) gas-to-electric conversion typical of a combined-cycle plant. If heavy-duty vehicles or hybrid passenger vehicles with their relatively high efficiency were to be substituted for the automobile shown, the combustion path would be close to the electric path.

EXPECTED GROWTH IN NATURAL GAS VEHICLE POPULATION

Several forecasts have been made for the growth in medium- and heavy-duty vehicles using NG. Figure 5-9 compares several forecasts of NG penetration in Class 8 trucks in the near term. Key factors influencing the decision to purchase a NG vehicle are the fuel cost savings, initial cost premium for the vehicle, and ready access to refueling facilities.

The National Petroleum Council (NPC) study produced a longer-term projection to 2050 (Figure 5-10), which estimated that about 40 percent of Class 7 and Class 8 vehicles would be fueled by NG by 2050.

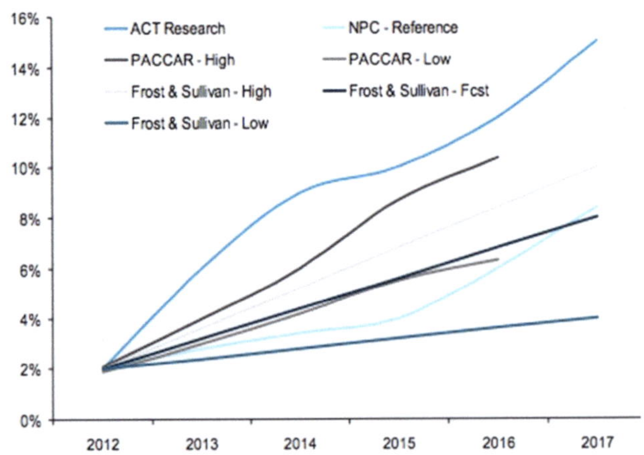

FIGURE 5-9 Near-term Class 8 natural gas penetration forecasts. SOURCE: Citi GPS (Global Perspectives & Solutions), 2013, *Energy 2020: Trucks Trains & Automobiles*. New York: Citigroup. June.

[9] Note that these are not full well-to-wheel estimates since they include neither leakage at the well nor tailpipe emissions from the CNGV.

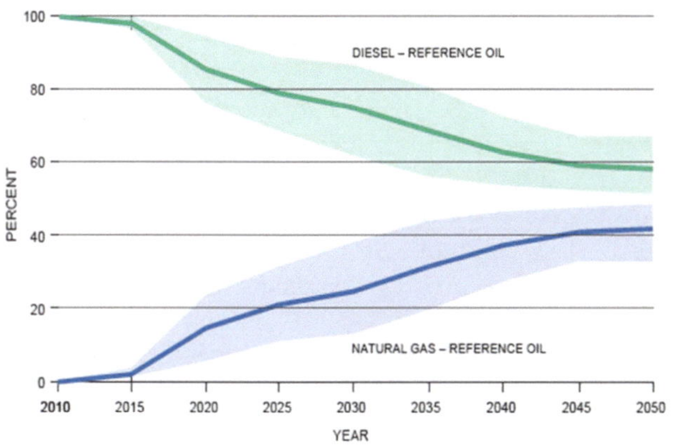

FIGURE 5-10 Shares of vehicles fueled by natural gas and by diesel to 2050. SOURCE: National Petroleum Council, 2012, *Advancing Technology for America's Transportation Future, Part One—Integrated Analyses*, pp. 3-17.

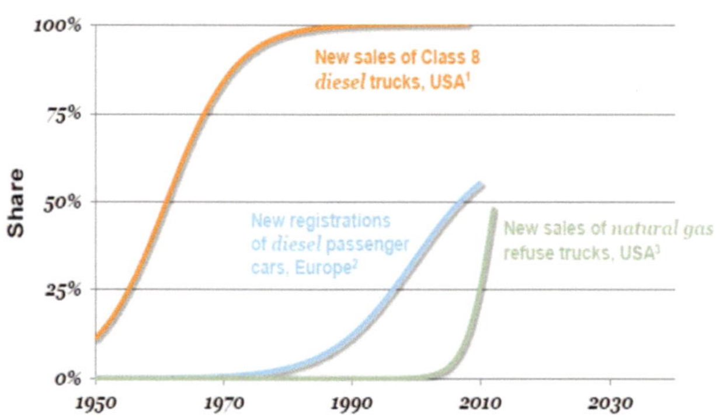

FIGURE 5-11 Historic rates of penetration for NG refuse trucks in the United States and diesel passenger cars in Europe. SOURCE: Data from (1) MacKay & Co. and Ward's Automotive Group, a division of Penton Media, Inc., (2) European Automotive Manufacturers' Association (ACEA), and (3) and (4) Westport.

The growth of refuse trucks fueled by NG is dramatic, as shown in Figure 5-11. Currently about half of new refuse trucks are NG-fueled, and that is expected to rise to 90 percent soon, in part because of local, state, and national incentives. However, refuse trucks collectively are not large consumers of fuel, so the greater opportunity for NG substitution is in Class 7 and Class 8 tractor-trailer rigs, which use about 20 times as much.

REGULATORY FRAMEWORK FOR NATURAL GAS ENGINES AND TRUCKS

New NG-fueled medium and heavy truck engines have been certified by their manufacturers to meet Environmental Protection Agency (EPA) criteria emission requirements and standards (NO_x, particulate matter [PM], etc.) for many years. The criteria emission tests and procedures for NG engines are the same as those used to certify gasoline- and diesel-fueled heavy engines. Most engines are certified using a transient engine test, referred to as the heavy-duty Federal Test Procedure (FTP), which is supplemented by additional tests (steady-state and not-to-exceed tests) that cover more engine operating conditions. Medium- and heavy-duty vehicle engines fueled by diesel fuel or NG are certified to a NO_x standard of 0.20 g/bhp-h. Certification data show similar NO_x emission levels for NG and diesel fuel. Evaporative testing is not required of NG vehicles. Some large diesel pickup trucks and vans are chassis-certified based on the vehicle test used to certify cars and light trucks. EPA specifies the properties of NG for the purpose of certification emission testing. It also allows use of alternative test fuel specifications if they represent fuel the engine would use in normal use, such as pipeline NG.

Greenhouse Gas Emission and Fuel Economy Standards for Engines

In August 2011 the federal government adopted CO_2 and fuel economy standards for new medium- and heavy-duty trucks and engines. The CO_2 emission standards for new engines vary by engine type and duty cycle. If the engine is gasoline fueled or uses a spark plug and operates like a theoretical Otto combustion cycle, it is considered a SI engine. A CI engine is defined as an engine that is not a SI engine.

The CO_2 emission standard for a new SI engine is 627 g/bhp-hr, beginning with model year (MY) 2016. For CI engines, the CO_2 standard depends on the engine class (light-, medium-, or heavy-duty) and on what type of vehicle the engine is used in (tractor or vocational), and it begins with 2014 models. The CO_2 emission standards for CI engines are shown in Table 5-5. The 2017 standards represent a 9 to 23 percent reduction in CO_2 compared to a 2010 baseline.

Small businesses (fewer than 1,000 employees for truck manufacturers and 750 employees for engine manufacturers) are exempt from the GHG and fuel consumption standards, although NHTSA and EPA state they plan on reexamining this in the next rulemaking.

TABLE 5-5 Engine Emission Standards for New CI Engines (grams per brake horsepower-hour)

		Engine CO_2 Emission Standard	
Vehicle Type	HD Engine Class	2014	2017
Tractor, Classes 7 and 8	Medium	502	487
	Heavy	475	460
Vocational	Light	600	576
	Medium	600	576
	Heavy	567	555

Methane and nitrous oxide engine emission standards of 0.10 g/bhp-hr each also apply to all engines regardless of fuel. Through 2016 models, CO_2 emission credits can offset methane (and nitrous oxide) emissions to provide more flexibility in meeting these additional standards, and the flexibility may help NG engines with high methane emissions comply.

The test cycle for measuring CO_2 emissions from engines used in Class 7 and Class 8 tractors is the supplemental emission test (SET). For engines used in vocational vehicles, the test cycle is the heavy-duty transient engine test procedure.

Fuel economy standards for new engines have been adopted by NHTSA. Compliance with the fuel economy standard is determined from the CO_2 emissions measured by the appropriate emission test. Thus an engine that meets the CO_2 standard also meets the fuel economy standard, and vice versa. The fuel economy standard for diesel engines does not begin until MY2017. SI engine fuel economy standards begin in 2016.

NG Engines

NG engines have to meet the same emission standards as gasoline- or diesel-fueled engines. NG engines are classified for criteria emissions based on the combustion cycle of the engine the NG engine is derived from. All larger NG engines (over 7 L) are currently derived from diesel CI engines even if they are converted to SI, and they are treated by EPA as diesel engines for criteria emission certification. EPA also requires consistent categorization for both criteria and CO_2 emission compliance. Thus, under current EPA requirements, larger heavy-duty NG engines, even if spark-ignited, will have to comply with the CI CO_2 standards shown in Table 5-5 beginning with the 2014 models. NG engines derived from gasoline engines are subject to the numerically higher SI CO_2 standard beginning with the 2016 models.

Dual-fuel engines and vehicles (such as NG and diesel) must be tested for CO_2 emissions on each fuel. Through 2015 the test results may be averaged; after 2015, the manufacturer must supply data on actual fuel use of each fuel for the type of vehicle, and the test results are weighted appropriately. The combined test results must comply with the standards. Criteria emission standards must be met on both fuels. However, separate evaporative testing of the CNG system is not required.

The CO_2 and fuel economy regulations change the engine classification definitions in a manner that may one day affect NG engines. A literal reading of the new definitions suggests that a SI engine derived from a diesel engine would be classified as a SI engine and would have to meet the SI engine CO_2 standard, which is less stringent than the diesel engine CO_2 standard it must meet in 2014. EPA has indicated that it may address this in a future rulemaking, but for now an engine manufacturer may choose to be classified based on either the new or the old definition.

Emission and Fuel Economy Standards for Complete Trucks

EPA and NHTSA have also adopted GHG and fuel economy standards for trucks, with compliance units of grams/ton-mile and gallons/1,000 ton-miles, respectively. The standards vary by truck type (tractor or vocational), vehicle class, and roof height of the tractors. These vehicle standards include no provision unique to NG engines. The vehicle CO_2 standards start in 2014, fuel economy standards start in 2016, and both become more stringent and aligned in 2017. Fuel economy and CO_2 fleet average standards were also adopted for large pickups and vans, which are chassis-certified. They are patterned after the adopted car and light truck CO_2 standards, and their numerical values vary by vehicle attributes such as size.

Heavy-duty NG engines are not expected to require significant design changes to meet the current GHG and fuel economy standards. CO_2 emissions from NG engines are typically 5 to 20 percent lower than those from comparable gasoline or diesel engines, and this is about the same reduction required by the current standards. The flexibility of using CO_2 reduction to offset methane emissions that are higher than the standard, as discussed in the section on GHG emissions and fuel economy standards for engines, suggests the methane standard will not be a barrier to compliance for NG engines. While NG leakage from a vehicle fuel system is technically possible, manufacturers and operators have great incentive to prevent this because it would represent a significant safety concern as well as a reduction in range.

Fuel consumption certification at the vehicle level for the new EPA-NHTSA standards is accomplished by use of the Greenhouse gas Emissions Model (GEM), a vehicle simulation tool with standard engine maps[10] as described in Chapter 3 of this report. At present, there is no engine map in GEM for NG engines, so NG-powered vehicles are included in an alternative certification process under the Advanced Technology and Innovative Technologies categories in the rule. Since NG-powered vehicles are expected to become significantly more prevalent in the coming years, future regulatory changes should consider modifying GEM to incorporate them. Using the procedures for Advanced Technology credits may be replaced by providing a standardized NG engine model (SI, stoichiometric, turbocharged) in addition to a diesel engine model. Otherwise, it could be multiplied by a factor relating it to a diesel engine. Although some NG engine technologies other than SI may be employed and may offer higher efficiency, the SI model should serve as a baseline owing to its current and likely future market penetration.

Several Class 2b pickup and van producers have questioned the ability of gasoline engines to comply with the stringent new rules. Should these gasoline engines be

[10] Engine maps are tables of engine input and output values determined from calibration tests. They represent how an engine will operate under specific conditions.

dropped from production (and probably replaced with diesel engines), the availability of Class 2b NG vehicles would be impacted because the NG engines with displacements of less than 7 L are generally derived from gasoline engines. EPA remains confident that technologies developed primarily for light-duty engines will allow gasoline engines to remain viable in this class. This issue should be monitored to determine if the availability of NG engines derived from gasoline engines will be impacted, and if this has an effect on achieving the maximum feasible reduction in fuel use and GHG emissions.

In summary, the current regulations accommodate NG engines and provide a pathway to their certification. Certifying the fuel economy of dedicated NG vehicles would be more straightforward if NG engine maps were added to the GEM vehicle simulation.

FINDINGS AND RECOMMENDATIONS

Finding: Most evidence points to a long-term price advantage and abundant supply for NG. Market risks include findings on leakage and environmental issues with fracking.

Finding: NG is a low-carbon fuel with a GHG advantage over diesel and gasoline that varies in magnitude with engine technology and duty cycles. Being a large domestic resource, NG use contributes to national energy security. The use of NG contributes to the NHTSA energy mission and EPA's goal of reducing climate impacts, unless additional findings of methane leakage alter this vision.

Finding: NG is also an economical fuel for many applications in medium- and heavy-duty vehicles. By 2025, 20 percent of new trucks sold could be fueled by NG, a very fast penetration rate from the current level. However, the regulatory approach and standards themselves are in flux, which could potentially disrupt normal market dynamics.

Recommendation 5.1: More studies and data are needed to determine the well-to-tank GHG emissions of NG vehicles, since current estimates vary significantly regarding quantification of emissions leakage of methane. EPA and NHTSA should assemble a best estimate of well-to-tank GHG emissions to be used as a context for developing future rulemakings.

Finding: With regard to the option of NG as a broad strategy for GHG reduction, NG engines are well-developed and can be readily employed in medium- and heavy-duty vehicles, especially with introduction of a 12 L engine for Class 8 trucks. There are areas of technology improvement for engine efficiency and maintenance cost reduction. On-vehicle storage of NG has high initial cost, economically restricting NG usage to applications with high fuel consumption.

The load-specific CO_2 emissions from NG engines are generally 5-20 percent less than those from comparable diesel engines. NG's inherent GHG benefit by virtue of its low carbon content (~28 percent) is partially negated by lower efficiency in currently available engines and the higher GHG impact of methane emissions. If the regulations require the same CO_2 levels for trucks using either diesel or NG, then it is conceivable a NG truck could meet the standards with fewer advanced fuel economy technologies than a diesel counterpart. In contrast, a NG leakage correction to GHG impact could negate the inherent tailpipe CO_2 advantage.

Recommendation 5.2: NHTSA and EPA should develop a separate standard for NG vehicles as is presently the case for diesel- and gasoline-fueled vehicles. Factors the Agencies should consider in setting the standard include the maximum feasible ability of NG engines to achieve reductions in GHG emissions and fuel consumption, the uncertainties involved with the various alternatives, the impact of duty cycles on the ability to comply with the vehicle standards, the cost of the technology, and rapid growth of the market for NG engines and vehicles. This may require additional focused studies.

Recommendation 5.3: Owing to the economics-driven rapid adoption of NG, it is urgent to develop an optimum solution in Phase II Rule standards for both GHG emissions and fuel consumption (as well as criteria emissions)—that is, a standard that will accommodate this fuel without artificially disrupting prevailing commercial transportation business models. Such standards need to be expedited to provide certainty to consumers and product developers alike, even as current business activities accelerate. As a specific example, the GEM certification tools need to include NG engine maps to more accurately quantify the emissions and fuel economy of NG vehicles.

Recommendation 5.4: To benefit fully from the GHG and petroleum displacement potential of NG, government and the private sector should support further technical improvements in engine efficiency and operating costs, reduction of storage costs, and emission controls (as is done for diesel engines). NHTSA and EPA should also evaluate the need for and benefits and costs of an in-use NG fuel specification for motor vehicle use.

Finding: The NG vehicle refueling infrastructure is growing, but the number of retail outlets for fueling medium- and heavy-duty vehicles is still well short of what is needed.

REFERENCES

AGA (American Gas Association). Undated. Available at http://www.aga.org/Kc/aboutnaturalgas/consumerinfo/Pages/NGDeliverySystemFacts.aspx. Accessed February 2014.

AFDC (Alternative Fuels Data Center), Department of Energy. Undated. Available at http://www.afdc.energy.gov/fuels/natural_gas.html. Accessed December 2013.

ANGA (American Natural Gas Association). U.S. and Canadian Natural Gas Vehicle Market Analysis. Undated, prepared by TIAX. Available at http://anga.us/media/content/F7D3861D-9ADE-7964-0C27B6F29D0A662B/files/Comparative%20and%20Scenario%20Analysis1.pdf.

ANGA. Gas Vehicle Market Analysis-LNG Infrastructure: Final Report. Undated, prepared by TIAX. Available at http://anga.us/media/content/F7D3861D-9ADE-7964-0C27B6F29D0A662B/files/LNG%20Infrastructure.pdf.

Aldrich, B. 1995. *ABC's of Afv's: A Guide to Alternative Fuel Vehicles.* Darby, Pa.: Diane Publishing Company.

Allen, D.T., V.M. Torres, et al. 2013. Measurements of methane emissions at natural gas production sites in the United States. *PNAS*, October 2013. Available at http://www.pnas.org/content/110/44/17768.abstract.

Celanese. 2013. Global Advocacy. Available at http://www.celanesetcx.com/en/global-advocacy. Accessed November 30, 2013.

Clay, Kathryn. 2013. "The Drive Natural Gas Initiative." Oral presentation at the SAE International Natural Gas Symposium. Greenville, S.C., March 12.

Curran, S., et al. 2013. *Well-to-Wheels Analysis of Direct and Indirect Use of Natural Gas in Passenger Vehicles.* Presented at SAE 2013 Fuels, Lubricants, and Aftertreatment Symposium: Achieving Fuel Economy and GHG Targets, November 18-21, 2013. (Article in press in *Energy*)

Energy Information Administration (EIA). 2013a. *Annual Energy Outlook 2014: Early Release.* Available at http://www.eia.gov/forecasts/aeo/er/index.cfm. Washington, D.C.: Department of Energy.

EIA. 2013b. Frequently Asked Questions.. Available at http://www.eia.gov/tools/faqs/faq.cfm?id=427&t=3. Accessed November 2013.

EIA. 2013c. *Annual Energy Outlook 2013.* Washington, D.C.: Department of Energy.

Engine Manufacturers Association (EMA). 2010. Letter to California Air Resources Board regarding ARB CNG Specification Revisions, June 3.

Greszler, A. 2011. *Opportunities for Heavy Vehicle Efficiency Improvements.* 2011 Biennial Conference: Rethinking Energy and Climate Strategies for Transportation. Asilomar, Calif. August 31.

Howarth, R.W., R. Santoro, and A. Ingraffea. 2011. Methane and greenhouse-gas footprint of natural gas from shale formations. *Climatic Change*, doi:10.1007/s10584-011-0061-5.

Kamel et al. 2002. An emission and performance comparison of the natural gas: Cummins Westport Inc. C-Gas Plus versus diesel in heavy-duty trucks. SAE 2002-01-2737.

Kauling, Dick. 2013. "US Light Duty Natural Gas Vehicles—Role of Customer Requirements and Standards in Technology Development." SAE Natural Gas Symposium. Greenville, S.C. March 13.

Krupnick, A.J. 2010. "Energy, Greenhouse Gas and Economic Implications of Natural Gas Trucks." Backgrounder, Resources for the Future. June.

NACFE (North American Council for Freight Efficiency and Cascade Sierra Solutions). 2013. *Barriers to the Increased Adoption of Fuel Efficiency Technologies in the North American On-Road Freight Sector*. July.

National Petroleum Council. 2012. *Advancing Technology for America's Transportation Future.* Washington, D.C. Available at: http://www.npc.org/.

National Research Council (NRC). 2010. *Technologies and Approaches to Reducing the Fuel Consumption of Medium- and Heavy-Duty Vehicles.* Washington, D.C.: The National Academies Press.

NRC. 2013. *Transitions to Alternative Vehicles and Fuels.* Washington, D.C.: The National Academies Press.

Pacific Northwest National Laboratory (PNNL). 2010. *Issues Affecting Adoption of Natural Gas Fuel in Light- and Heavy-Duty Vehicles.* PNNL-19745. Richland, Wash.: PNNL. September.

TIAX. 2009. *Assessment of Fuel Economy Technologies for Medium- and Heavy-Duty Vehicles.* Report prepared for the National Academy of Sciences by TIAX LLC. Cupertino, Calif. July 31.

University of Michigan Transportation Research Institute (UMTRI). 2013. *Trucks Involved in Fatal Accidents,* 2006-2010.

Weeks, B. 2013. Technology Trends. Oral presentation at the SAE International Natural Gas Symposium, Greenville, S.C. March 12.

6

Review of Options to Reduce Energy Use of Trailers

This chapter addresses the opportunities to reduce the energy consumed by Class 8 tractors pulling, particularly, van trailers. Following some background information, three government programs that deal with tractor-trailer fuel consumption are summarized. Next, the technologies associated with tractor and trailer aerodynamics as well as tires for both components are discussed. The contribution to life-cycle costs of tire pressure monitoring (and maintenance) systems (TPMS) and greenhouse gas (GHG) emissions will also be considered. Finally, the findings and appropriate recommendations are presented.

Because the tractor and trailer act as a system, with each part affecting the energy use of the other, options to reduce energy use of the tractor are also briefly discussed. While tractors are built for the weight Classes of 8, 7, and 6, the most populous and versatile and the default industry workhorses are Class 8 tractors. Reduced tare weight is noted as a contributor to reduced energy consumption (or, alternatively, to marginally increased payload) and is not discussed further.

A fully loaded Class 8 tractor-trailer combination operating on the interstate at a constant 65 mph typically demands over 200 hp from the engine. This power demand is principally to drive the wheels at freeway speeds to overcome aerodynamic drag and tire rolling resistance. The remaining power demand, in the absence of grade or headwinds, is to overcome drivetrain friction and to power auxiliary devices. Table 6-1 details these demands.

Class 8 tractor-trailers account for 60 percent of the fuel used by all on-road heavy-duty trucks (ICCT, 2013). The disproportionate fuel use notwithstanding, Class 8 tractor-trailers are relatively small in number because of the just-mentioned high power demands at freeway speeds (65 mph) and the high annual mileages accumulated by these vehicles (a median of about 100,000). By comparison, Class 3 to Class 6 fully loaded delivery trucks require less than a third of the power to operate at a constant urban speed of 40 mph, and they each accumulate fewer miles per year (a median of about 40,000) (NRC, 2010, Tables 2-1 and 5-2). Therefore, straight trucks with these predominately urban duty cycles will not be further considered in this chapter.

In addition to trailers towed by tractors, some trailers are also transported by rail. "Intermodal transport" refers to the movement of goods by more than one mode on a single journey (Corbett and Winebrake, 2007; Winebrake et al., 2008). Commonly, intermodal transport combines a truck mode with either ship or rail to improve shipping efficiency, reduce costs, or achieve some other desirable performance attribute. Because rail and ship are significantly less energy-intensive than truck, incentivizing the movement of goods from truck to rail or ship is one way to improve the overall efficiency of the freight transportation system (NRC, 2010, p. 175).

Containers are transported at each end of their route by truck tractors. These final segments are typically much shorter than the total journey of the container. The container is on- and off-loaded to a chassis, which completes the trailer configuration (sometimes standard van trailers are also rail transported). When the notion of adding trailer aerodynamic devices is considered later in this chapter, the potential interference of those devices with container handling must be considered.

TABLE 6-1 Operational Power Demands from Class 8 Tractor with Sleeper Cab-Van Trailer at 65 mph on a Level Road and Having a Gross Vehicle Weight (GVW) of 80,000 lb

Operating Load	Power Consumed (hp)	Power Consumed (%)
Aerodynamic	114	53
Rolling resistance	68	32
Auxiliaries	20	9
Drivetrain	12	6
Total	214	100

SOURCE: NRC, 2010, Table 5-4.

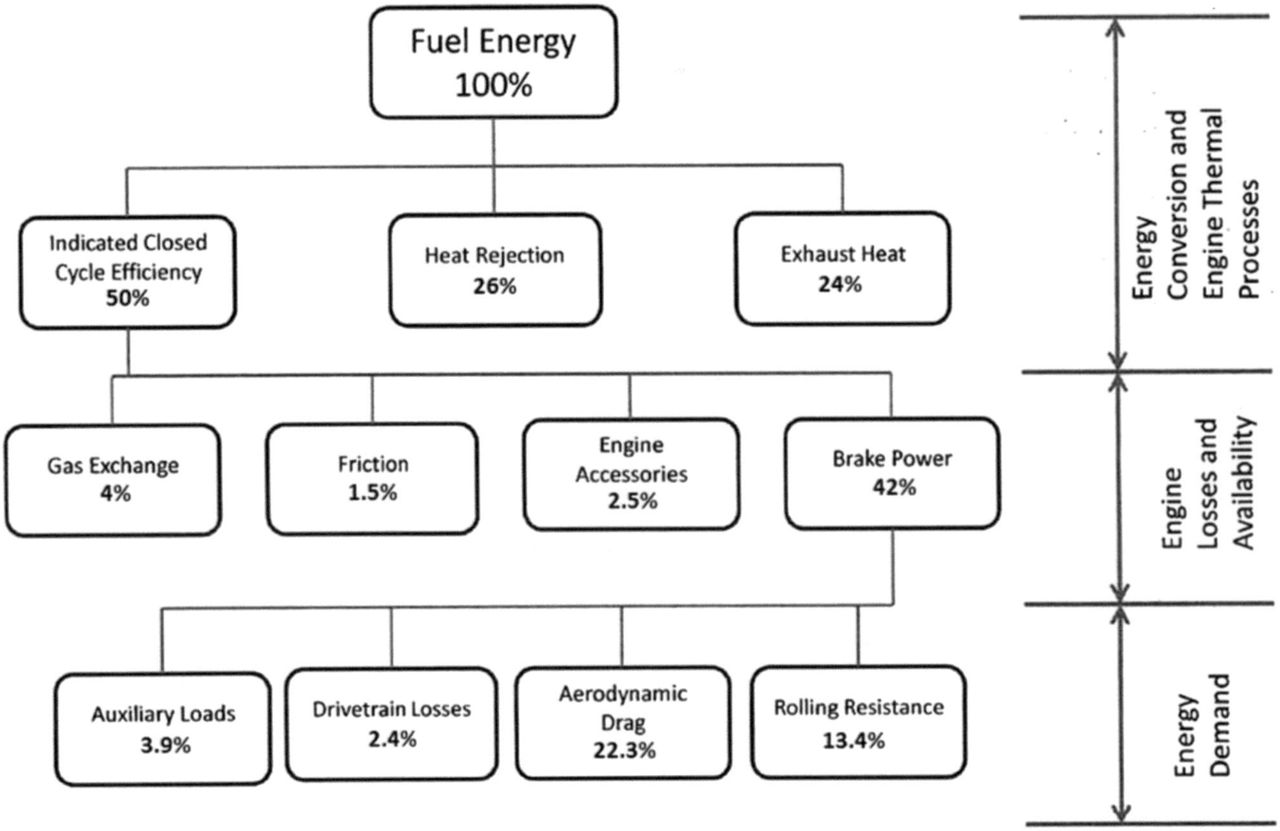

FIGURE 6-1 Energy balance of a fully loaded Class 8 tractor-trailer on a level road at 65 mph. SOURCE: NRC (2010), Figure 5-1.

BACKGROUND

Current Tractor-Trailer Energy Balance

Efficiency of Class 8 tractor-trailers can also be improved by reducing energy losses in the engine processes. As shown in Figure 6-1, half the fuel energy available to the engine is typically lost to heat. Another 8 percent of the fuel energy is lost to overcoming pressure differentials and friction and to power accessories. The remaining energy available, 42 percent, is the operational power demand from the engine and drivetrain, as summarized in Table 6-1.

Note that not all Class 8 combination trucks operate primarily at high interstate speeds, nor do they all carry high gross weights. Indeed, many operate in shorter regional haul or mixed highway and suburban duties and often carry partial loads. Still others may operate substantially in suburban and urban areas, often with frequent stops. Often these shorter haul duties utilize day cab tractors, typically returning to central dispatch for overnight domicile. These descriptions serve to clarify that there is a continuum of duties in the tractor-trailer universe, and the benefit of technologies that reduce fuel use will vary widely depending on speed, the load being carried, and the mileage accumulated.

Aerodynamics and Tire Rolling Resistance of the Tractor-Trailer

The power required to propel a vehicle at any moment in time is customarily presented as a "road load equation." This power equation has four terms to describe tire rolling resistance, aerodynamic drag, acceleration, and grade effects:

$$P_{RL} = mgC_{rr}V\cos(\theta) + 0.5C_dA\rho_aV^3 + mV(dV/dt) + mg\sin(\theta)V$$

where P_{RL} is road load power, mg is vehicle weight, C_{rr} is tire rolling resistance, C_d is a drag coefficient based on the entire vehicle, A is the frontal area, ρ_a is the air density, V is the vehicle velocity, m is vehicle mass, t is time, and θ is the road gradient (uphill positive). Neither C_d nor C_{rr} need be constant with respect to speed and are not treated as constant in the better simulations. The term C_dA should not be separated as it arguably represents a fundamental characteristic of the vehicle for which it has been determined.

The power required to overcome aerodynamic drag is proportional to the cube of forward velocity. This illustrates the important influence of vehicle speed on horsepower demand and the fuel consumption needed to overcome aerodynamic drag. The power required to overcome tire rolling resistance

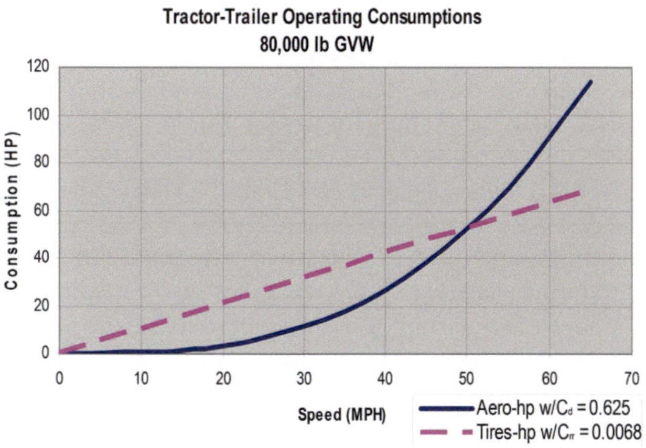

FIGURE 6-2 Aerodynamics and tire power consumption for a late-model tractor-trailer combination with GVW of 80,000 lb. SOURCE: NRC, 2012, Figure 5-1.

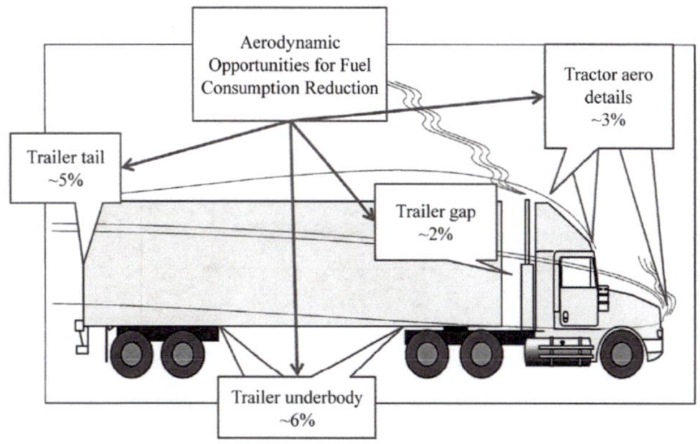

FIGURE 6-3 Tractor-trailer combination truck illustrating regions of potential fuel consumption reduction; combined C_d base of 0.625. SOURCE: NRC, 2010, Figure 5-8.

is proportional to the forward velocity. These relationships are shown in Figure 6-2.

Notice that these two power demands are equal at about 50 mph, while rolling resistance power consumption at 36 mph is about twice that required to overcome aerodynamic drag. This comparison is for a fully loaded circa MY2007 tractor[1] and a trailer that does not incorporate aerodynamic devices. An empty or partially loaded trailer would have less rolling resistance because the rolling resistance force is also proportional to the weight on the tires.

Aerodynamics of the Combined Tractor-Trailer

There are four regions of the tractor-van trailer combination truck that are amenable to aerodynamic design improvements. These regions include the various tractor details, the tractor-trailer gap, the trailer underbody, and the trailer tail (NRC, 2010, p. 96). These are illustrated in Figure 6-3, along with the estimated fuel consumption reductions that might be realized for trailers with aerodynamic devices and present-generation sleeper tractors.

Tractor Aerodynamics

The heavy truck industry began purposeful tractor aerodynamic improvements through hood and fender styling changes, plus aerodynamic bumpers and fuel tank fairings and by moving the externally mounted air cleaner canister under the hood. These changes were led by the introduction of Kenworth's T600 tractor model in 1985. That game-changing introduction spurred the entire industry to accelerate aerodynamic tractor styling, as customers began to measure fuel consumption reduction effects.

The current aerodynamic features for tractors are identified in Figure 6-4 and can be seen in the photos in Figure 6-5, which illustrates the most significant aerodynamic differences and similarities between aerodynamic and nonaerodynamic tractors. Panel (a) shows a MY2013 aerodynamic high-roof sleeper cab tractor, equipped with aerodynamic hood, fenders, bumper and mirrors, side fairings, and an integrated roof fairing with cab extenders that help reduce the turbulent area between the cab and the front of the trailer. Panel (b) depicts a day cab tractor equipped with only a simple roof air deflector and no side fairings, yet common cab, hood, fenders, bumper, and mirrors. As the truck industry rarely reports C_d values, such figures are not available for the tractors shown. Indeed, tractor-only C_d values are of limited value, since the combined drag of the tractor and trailer is most significant to the truck's fuel consumption.

Tractor manufacturers introduce design modifications periodically, and their advertisements often claim aero-related performance improvements (including for engine performance). Indeed, these often are declared by the manufacturers to be major, competitive, purchase-worthy steps forward. It is anticipated that the Department of Energy (DOE) SuperTruck projects[2] will generate significant aero-

[1] Model years vary significantly among medium- and heavy-duty vehicle (MHDV) manufacturers, so for the sake of simplicity and uniformity, the calendar year is often used as the rough approximation for model year.

[2] SuperTruck is a major initiative of DOE's 21st Century Truck Partnership and is supported by three other federal agencies in cooperation with fifteen industrial partners. The latter include all six U.S. heavy-duty truck manufacturers and many heavy-duty engine and powertrain system manufacturers in order to accelerate technology development and provide focus for R&D efforts. The four industry-led projects, spanning 2010 through 2016, were established to fund R&D and demonstration of full vehicle systems integrating a number of technologies into Class 8 heavy-duty, long-haul trucks. There are three major goals: (1) demonstrate a 50% increase in freight efficiency (measured in ton-miles per gallon) on a defined drive cycle with a 20% engine contribution; (2) demonstrate engine

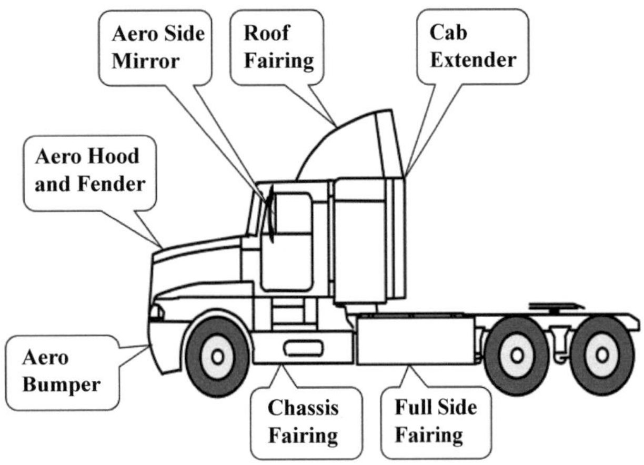

FIGURE 6-4 Sleeper tractor with aerodynamic features identified. SOURCE: NRC, 2010, Figure 5-5.

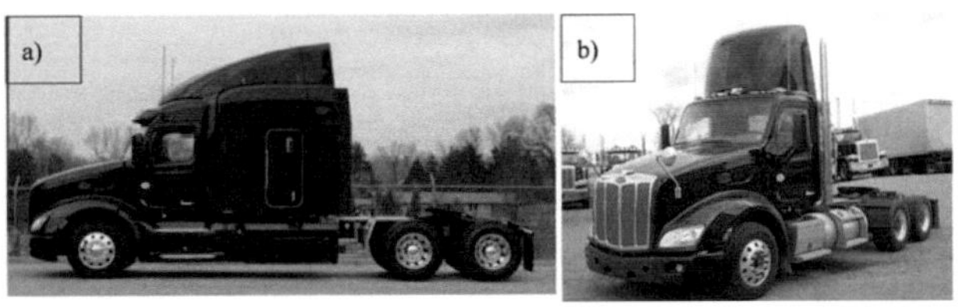

FIGURE 6-5 (a) Peterbilt aerodynamic High Sleeper Cab and (b) Day Cab model 579 tractors. Courtesy Peterbilt Motors Corp., 2013.

dynamic innovations when they conclude between 2014 and 2016. The Phase I Rule also provides a strong incentive to tractor manufacturers to continuously reduce fuel consumption.

Van Trailer Aerodynamics

Improvements to the aerodynamics of van trailers are influenced by customer demand for reduced fuel consumption, the Environmental Protection Agency's (EPA's) SmartWay voluntary program, and GHG regulations of the California Air Resources Board (CARB) (discussed in a later section). The current federal Phase I Regulations on fuel economy and GHGs apply only to engines and tractors and do not require trailer manufacturers to reduce the impact of their trailers on tractor fuel consumption.

The most common improvement to trailers is the addition of side skirts to improve underbody aerodynamics (see discussion in the section "Current Use of Aerodynamic Devices and Low Rolling Resistance Tires"). These may be added to new trailers or retrofitted to existing trailers. The CARB regulation includes a GHG reduction requirement that is most easily met with skirt retrofits, and this has created a burgeoning aftermarket for them. Other types of devices such as nose cones and rear fairings are used on a small fraction of trailers. EPA testing shows that most side skirts provide a 5 percent fuel saving at 65 mph (3 percent at 55 mph), as do most rear tail fairings (see Table 6-2). Other devices such as trailer front gap devices (see Figure 6-6), wheel covers and fairings, and vortex generators and flow tabs have been developed. Overall, the use of trailer aerodynamic devices has grown significantly in the past few years. See the SmartWay discussion below that references devices verified or qualified in that program.

Tractor-Trailer Gap

The airstream leaving the tractor cab encounters the gap between tractor and trailer.[3] The gap is highly turbulent with air motion out of control and pressure further reduced. Yet behind an aerodynamic tractor, many measurements have identified this region as providing only a 0.5 to 3.5 percent opportunity for drag reduction (TIAX, 2009). An average performance of 1.3 percent was achieved with partial gap closures and 2.2 percent with full gap closures. Because the CARB regulation requires any trailer aero-performance improvement to reduce fuel consumption by at least 5 percent, which can be achieved by using skirts under the van trailer body, the typical partial gap closure devices on offer are utilized only infrequently.[4]

Tire Rolling Resistance

Nearly all heavy-truck tire manufacturers produce low-rolling-resistance (LRR) tires for all wheel positions on tractor trailers, and many of these have been performance verified by SmartWay. SmartWay requires meeting specific rolling resistance targets based on test data in order to be verified. Achieving the target values results in a 15 percent reduction in rolling resistance, measured against a 2010 baseline tire. Best-in-class tires provide a 30 percent reduction in rolling resistance, indicating that still greater reductions in rolling resistance beyond the SmartWay targets are possible (76 Fed. Reg. 57207). Improvements in rolling resistance have been achieved with new tread compounds, stabiliza-

50% brake thermal efficiency (BTE) at a level road load of 65,000 lb, 65 mph; and (3) conduct technology development and scoping for a path to 55% BTE. SOURCE: Ken Howden, DOE, "21st Century Truck Partnership and SuperTruck Initiative," Presentation to the committee, March 29, 2013.

[3] Freight Wing, Inc. "On Aerodynamics: The Effect of Aerodynamics on Tractor Trailers." Available at http://www.freightwing.com/on-aerodynamics.php. Accessed December 18, 2013.

[4] Sean Graham, Freight Wing Incorporated, personal communication with Chuck Salter, committee member, August 26, 2013.

TABLE 6-2 SmartWay-Verified Devices for Heavy-Duty Tractor/Trailers

Device	Fuel Economy (mpg) increase (%)	Number verified
Advanced trailer side skirts	≥5	37
Trailer side skirts	≥4	16
Advanced trailer end fairings	≥5	9
Trailer boat tails	≥1	9
Trailer gap reducers	≥1	4
LRR tires (new)	≥3	41 brands
LRR retreads	≥3	7 brands
Certified tractors	Specifies design elements	19 models
Certified 53+ foot trailers (new)	Specifies aerodynamic configurations	8 manufacturers

NOTE: mpg, miles per gallon.
SOURCE: www.epa.gov/smartway/forpartners/technology.htm. Accessed August 29, 2013.

tion of the tread block, and stiffer shoulders to reduce tread deformation. (Note that a 15 percent reduction in C_{rr} for the entire truck translates into a reduction of about 3 percent in fuel consumption.)

Manufacturers have also introduced wide-base single tires (WBST), many of which feature lower rolling resistance than most dual tire sets and most often produce best in class C_{rr}. The WBST must be mounted on a special wheel and axle end, which increases cost but has the additional benefit of reducing wheel weight and thus increasing payload capacity.

FIGURE 6-6 Trailer front gap fairing. SOURCE: Freight Wing. Available at http://www.freightwing.com/gap_fairing.php. Accessed November 14, 2013.

Tread life and retreadability also play into a carrier's analysis of how to achieve the lowest costs over a tire's life cycle.

Following is a treatise on the factors influencing successful WBST use, offered by an experienced truck industry executive (though not a tire manufacturer):

> These experiences are primarily those for over the road, tractor-trailer fleets. Wide Base Single Tires (WBST) make up less than 5% of the commercial truck tire market, but their use is increasing due to fleets' desires to reduce fuel consumption and greenhouse gas emissions. Since 2000 when first introduced by Michelin, many fleets have tried and adopted them while others have tried and abandoned them for several reasons.
>
> Satisfied fleets feel the savings in fuel (between 2 and 5%) greatly offsets any disadvantage in tread wear and tire life. These fleets often utilize automatic tire inflation systems to keep the tires running when trailers encounter road hazards that puncture the casing so that they can get home for repairs.
>
> Fleets that abandoned them usually did so because they did not provide necessary maintenance and experienced high numbers of emergency breakdowns. Some fleets experienced extremely long breakdown periods (e.g., double the 2 hours for standard tires), often due to unavailability of wide base spare tires especially in certain areas of the country.
>
> WBSTs are very sensitive to both under and over inflation. The Tire & Wheel (S.2) Study Group of the Technology & Maintenance Council found fleets experiencing both shoulders wearing away usually were over-inflating these tires. Accordingly several tire manufacturers now recommend 100 psi as the optimal pressure to avoid this condition. Tire pressure systems (TPS) are clearly a robust solution to maintaining proper tire pressure for WBSTs, as well as steer and dual tires. See Subsection 6.4.1 below for an extended discussion on the value of TPS. WBSTs are also more sensitive to free rolling wear which is aggravated by lightly loaded trailers. Negative camber in the axle can cause the shoulder to wear prematurely which can be a big issue for fleets hauling heavier loads.
>
> Irregular wear negatively impacts tread mileage. In addition WBSTs usually come with slightly less tread depth than standard dual tires. Some fleets report getting 50-60% of the tread mileage they experience with duals. Yet other fleets achieve a tread mileage nearly equal to dual tires, through inflation maintenance and proper alignment. Improvements in compounding and tread design find WBST tread life now approaching the same rate as dual tires.
>
> Early retreadability of WBSTs was unsuccessful due to their higher unit loads causing faster casing fatigue. Now, most fleets usually get 1 retread from a wide base casing. This is certainly a short fall for fleets that routinely get 2 retreads on standard dual tires. But again, the savings in fuel economy can make up for this short fall. Some of the problems attrib-

uted to the reduction in retreads fall on retreaders, some of which are still climbing their learning curve.[5]

Tire manufacturers have variously reported the following for WBST tread life:

- With wide-base C_{rr} 15 percent below duals (the most popular wide base), it is 67 percent or more of tread life of dual tires.
- WBST tread life is nearly equal to the tread life of duals manufactured to the same C_{rr}.

It is noted from the foregoing that while the use of WBST LRR tires results in better fuel economy, the shorter tread life of these tires may lead to an increase in the number of newly manufactured tires and retreads, thereby generating additional life-cycle GHG emissions. The balance between these contrary effects is not known.

GOVERNMENT PROGRAMS THAT INFLUENCE TRACTOR-TRAILER FUEL CONSUMPTION

SmartWay

SmartWay was formed in 2004 as a collaboration of the EPA and the goods movement industry. The objective of SmartWay is improving efficiency and reducing fuel consumption and pollution from the movement of freight across the supply chain. Currently its focus is on-road trucking, which carries the majority of the nation's freight. The completely voluntary SmartWay Partnership grew more than 10-fold between 2006 and 2012 to 3,000 partners.[6] Partners agree to provide data on their operations, which are input into standardized tools to produce a measurement of environmental efficiency, such as grams pollutant per ton-mile. These benchmark results are compared to similar categories of freight movement—for example, dry vans—and ranked in quintiles. Fleet operators and shippers use the results to improve their efficiency, identify green options, and achieve recognition.

SmartWay also includes a Technology Program that certifies the performance of technologies, equipment, and strategies that save fuel and reduce emissions. This process increases the certainty for potential users of equipment and strategies that fuel savings will occur and a reasonable return on investment can be achieved. The technology program verifies the fuel savings of new tractors, new trailers, aerodynamic devices retrofitted to trailers, and LRR tires. New tractors are verified by design category based on their use of an integrated high roof sleeper cab fairing, cab side extenders, fuel tank fairings, aerodynamic bumpers and mirrors, LRR tires, or a device that provides 8 hours of idle-free power and cabin conditioning. No testing is required.

New dry van (nonrefrigerated) trailers 53 feet or longer ("53+ ft") may also be verified. Verification is based on use of a combination of SmartWay-certified devices and tires that reduce trailer drag and rolling resistance and provide at least a 6.5 percent reduction in fuel use relative to a baseline trailer. Typically this includes use of an advanced trailer skirt or, less frequently, an advanced trailer end fairing ("advanced" refers to a device verified to provide at least a 5 percent reduction in fuel use). Other combinations of verified devices are possible. Trailers must also use verified LRR tires. Aerodynamic devices designed to be retrofitted onto trailers must be tested to demonstrate their fuel reduction efficacy. The test involves comparing the fuel use of two identical trucks, one equipped with the drag reducing device(s) and driven on a dry, closed-loop test track in low wind conditions at no more than 65 mph and the other not so equipped . The procedure is specified as modified Society of Automotive Engineers (SAE) standard J1321. The results place the device into one of five verification categories: gap reducers, end fairings (1 percent or greater reduction in fuel use, each), side skirts (4 percent or greater), advanced skirts, or rear fairings (5 percent or greater, each). This test may also be used for other types of aerodynamic devices provided EPA establishes a new SmartWay verification category for the technology.[7]

To add some perspective to this test procedure's maximum speed, it is instructive to review the Freight Performance Measure study by the Federal Highway Administration of the five busiest interstate highways, which together account for nearly 25 percent of interstate freight vehicle miles traveled. Average truck speed deficits from posted limits are attributable to numerous causes, infrastructure design and capacity being the most common cause, plus terrain, weather, accidents, work zones, and time of travel (including operational strategy). That analysis showed an annual average for the four highest speed highways (of the five) in the range of 54 to 58 mph, road by road. The worst interstate was I5 at 50 mph, surprising few. Implied is that certain of the remaining interstates may embody somewhat higher averages, to be determined by a second study.

This suggests that test procedures evaluating the fuel savings of aerodynamic devices should include the variety of speeds experienced on real roads (FHWA, 2006).

LRR tires may be verified using the SAE J1269 or ISO 28580 test procedures, which involve measuring the steady-state rotational force required to turn the tire against a drum

[5] Peggy J. Fisher, President, TireStamp, Inc., "Low Profile Metric Wide Base Radial Tire Issues," personal communication to Chuck Salter, committee member, November 7, 2013.

[6] Sam Waltzer and Cheryl Bynum, EPA, "US National Approach to Reducing Freight Emissions," personal communication to Tom Cackette and Chuck Salter, committee members, July 24, 2013.

[7] Detailed requirements for obtaining SmartWay verification for tractors and trailers may be found at http://epa.gov/smartway/technology/designated-tractors-trailers.htm. Accessed August 29, 2013.

under specified conditions. To be SmartWay verified, the tire's rolling resistance coefficient must be less than EPA's target values, which vary depending on where the tire is positioned on the tractor/trailer. In general, LRR tires reduce fuel use by about 3 percent, with roughly half the reduction coming from the trailer. In June 2012, EPA also issued an interim protocol and procedures for measuring rolling resistance of retread truck tires.[8]

SmartWay-Verified Technologies

Only three of the eight trailer manufacturers with SmartWay-verified 53+ ft trailers highlight the use of side skirts or the SmartWay program through use of pictures and/or discussion on their websites. Two make no mention at all, and the rest offer skirts through an option list. Only one mentions the regulatory requirements of California. Most major commercial truck tire manufacturers discuss on their websites tire models that reduce fuel use and mention SmartWay-verified models in their tire selector or in brochures. Classes 7 and 8 tractor manufacturers all highlight aerodynamics and fuel efficiency of their trucks, but they continue to produce some Class 8 models that offer the "classic look," which does not utilize hood shapes and fuel tank covers that reduce aerodynamic drag. These are typically designed to haul flatbed trailers and tankers; one manufacturer reports that the classic design accounts for less than 5 percent of new tractor sales.

SmartWay also verifies idling reduction devices (which may be exempt from federal excise tax) and retrofit devices aimed at reducing smog-forming criteria emissions.

Improvements to SmartWay

EPA is undertaking studies to correlate the on-road performance of devices with the results of manufacturer verification testing. For example, an EPA contractor is conducting on-road fuel consumption testing pursuant to SAE J1321 of verified tires and aerodynamic devices to establish correlations with laboratory and track verification test results. One objective is refining current procedures used for verification.

Such attention to test procedure improvement is necessary and timely. During the committee discussions and data collection with tractor, trailer, and aerodynamic device manufacturing personnel, inadequacy of the required J1321 procedure was frequently cited.[9, 10] Significant issues include the following:

- Regulation demands more precision than a voluntary SmartWay program does. Fuel consumption reduction verification in the range 1 to 7 percent requires better than the ±1 percent reported in J1321.
- Supposedly identical tests run several months apart showed a 2 percentage point variation in fuel consumption reduction (first 4 percent then 6 percent).
- Results from an aero-device manufacturer's test could not be replicated within 2 percentage points by a trailer manufacturer.
- Results between test facilities were sometimes not precise within 2 to 3 percentage points.
- A single trailer sometimes performed very differently when towed with different SmartWay-verified sleeper tractors.

Improvement of the precision of SAE J1321 aerodynamics test procedure appears particularly important. Dissemination of these results to fleets and operators will increase confidence in the efficacy of fuel-saving devices.

Scale-model wind tunnel testing for trailers is also being conducted to establish correlation to on-road performance. Scale-model wind tunnel testing could reduce the cost of verification for trailer aerodynamic devices. SAE has developed recommended practice SAE J1252 for this testing. A wind tunnel procedure is the only accurate method of determining wind-average drag, by accounting for the effects of side winds. A test procedure that has no systematic process to account for yaw calls into question the value of devices that perform at a higher level in the presence of yawing wind. It will be key for EPA to establish a good correlation between wind tunnel tests and its modified J1321 on-road tests to help reduce the costs of development and verification for aero devices. It is significant to note that at least three of DOE's four SuperTruck project teams are utilizing either scale-model or full-size wind tunnels to provide quality analysis for the aero component of those projects.[11]

The EPA is also exploring the possibility of using computational fluid dynamics (CFD) to determine trailer aerodynamic drag for use in verification. CFD continues to be applied to full truck aero developments. Here again, three of four SuperTruck project teams report using these flow visualizing and quantifying tools to reduce analysis time and avoid prototype builds. EPA should consider processes to permit the results of these time- and cost-saving tools to satisfy performance verification requirements.

Lab testing of new and retread tires is being conducted to establish correlation to road testing and to potentially improve the current lab procedures. This may lead to

[8] Further information available at http: http://www.epa.gov/smartway/forpartners/technology.htm. Accessed February 13, 2014.

[9] Jeff Bennett, Utility Trailer Manufacturing Company, personal communication to Chuck Salter, committee member, August 29, 2013.

[10] Sean Graham, Freight Wing, Inc., personal communication to Chuck Salter, committee member, August 26, 2013.

[11] U.S. Department of Energy Vehicle Technologies Office, Annual Merit Review and Peer Evaluation Meeting. May 13-17, 2013. Available at http://www.annualmeritreview.energy.gov/. Accessed November 5, 2013.

improvements to the test procedures for verifying the efficacy of SmartWay-verified devices, trailers, and tires.

Finally, EPA is testing SmartWay idle reduction systems in a full-scale environmental chamber to better understand the energy load demand on the truck cab and explore a performance-based protocol for aero devices.[12]

EPA is considering inclusion of refrigerated 53+ ft van trailers and twin trailers (twin 28 ft pups) in SmartWay. It is also evaluating adding an "Elite" category for SmartWay trailers that has a higher fuel savings target, such as 10 percent. This could be met by use of multiple aerodynamic devices such as advanced side skirts and an advanced rear fairing (each providing a 5 percent reduction).

California Air Resources Board Regulation

In December 2008, CARB adopted a regulation to reduce GHG emissions from heavy-duty trucks by requiring new and in-use tractors and trailers to utilize aerodynamic designs and devices and LRR tires when operating in California. The regulation specifies that equipment used to comply with CARB's regulation must be verified through EPA's SmartWay program. The regulation applies only to 53+ ft van-type trailers (dry and refrigerated) and the Class 7 or 8 sleeper cab tractors that pull them. The regulatory requirements are summarized in Table 6-3.

Several salient points of the regulation are as follows:

- The regulation applies to the tractor-trailer operator, not the manufacturer of the tractor or trailer.
- The regulation applies to any affected tractor-trailer operating on a California road—that is, not just to those trucks based or licensed in California. Thus the regulation is expected to affect truck operators and owners based in other states that deliver products to and from California.
- The regulation requires tractors and trailers of MY 2011 and later to meet SmartWay requirements as follows: All 53+ ft van trailers regardless of what type of tractor is pulling them; all sleeper cab tractors if pulling a 53+ ft van trailer of any MY. In addition, non-sleeper-cab tractors must use LRR tires.

Because the regulation provides several options that can extend the final compliance date beyond 2013, some truck operators may use 2010 and earlier noncomplying trailers until as late as 2019 if certain conditions or interim milestones are met.

[12] Sam Waltzer and Cheryl Bynum, U.S. Environmental Protection Agency, "SmartWay Technology Program: Influencing Efficient Freight Movement into the Future," personal communication to Tom Cackette and Chuck Salter, NRC Committee on Technologies and Approaches to Reducing the Fuel Consumption of Medium- and Heavy-Duty Vehicles, Phase Two, July 24, 2013.

TABLE 6-3 Summary of CARB Regulatory Requirements

Type of Tractor and Trailer	Requirement and Timing
MY2011-2013 sleeper cab tractor	EPA SmartWay-certified.
MY2011-2013 day cab tractor	SmartWay-verified low rolling resistance tires.
MY2011-2013 model year 53+ van trailer	EPA SmartWay-certified. *or* Use SmartWay-verified aerodynamic device(s) that have been demonstrated to achieve at least a 5% fuel savings (4% for refrigerated trailers) *and* use SmartWay-verified LRR tires.
MY2010 and earlier tractors	Use SmartWay-verified LRR tires by 1/2013, both sleeper- and non-sleeper-cab tractors.
M 2010 and earlier trailers	Use SmartWay-verified device(s) by 1/2013 that have been demonstrated to achieve at least a 5% fuel savings (4% for a refrigerated trailers) *and* use SmartWay-verified LRR tires by 1/2017.

SOURCE: http://www.arb.ca.gov/msprog/truckstop/trailers/ttghg_ regorder.pdf and www.arb.ca.gov/regact/2013/hdghg2013/hdghg2013isor.pdf. Accessed November 18, 2013.

NHTSA and EPA Regulations

As noted in the introduction to Chapter 1, EPA and the National Highway Transportation Safety Board promulgated standards on GHG emissions and fuel consumption in 2011 (EPA and NHTSA, 2011). The two agencies declined at that time to regulate trailers. However, the GHG emissions model (GEM) used to certify engine and vehicle compliance includes the performance of a default van trailer for line-haul trucks. EPA indicates it only has the authority to set requirements for new trailers, whereas the CARB regulation requires existing trailers to install aerodynamic devices and existing tractors and trailers to use LRR tires.[13]

Methods for Aerodynamic Performance Evaluation

Regulations require reliable means of evaluation or testing to verify compliance. The value of compliance tests depends on ease of use, accuracy and precision of results, replicative fidelity, availability of test facilities, and relevance of the compliance test to real-world applications.

As noted in an earlier section, "Improvements to SmartWay," both EPA and several users of the current SmartWay test procedures acknowledged inadequacies in the fidelity

[13] William Charmley, U.S. Environmental Protection Agency, personal communication to Tom Cackettte, committee member, August 2013.

and especially in the precision of results when using the modified TMC/SAE J1321 Type II. While this recommended practice may appear to be the default procedure for a future regulation, the various compliance tests and procedures need to be evaluated to determine which one(s) are best suited to the particulars of the new regulation.

There are numerous SAE-recommended practice protocols available to evaluate the aerodynamic performance of commercial vehicles and aerodynamic devices. Three in particular address full-scale commercial vehicles in open air conditions either on the highway or at the test track.

SAE J1321 was first published in 1986. It consists of test and analysis methods to determine the change in fuel consumption for trucks and buses with GVW rating (GVWR) of more than 10,000 lb. To improve the precision of the results for use in SmartWay, EPA developed 22 additional provisions for SAE J1321. Examples of the provisions include requirements that the test must be conducted on a test track, not a roadway, with a grade change on the test track not greater than 2 degrees and the test facility not more than 4,000 ft above sea level. The latest revision occurred in 2012 as TMC/SAE J1321 Type II. This recommended practice may be used to compare entire vehicles and for easy-to-change components. The revised recommended practice specifically states that "test results that do not rigorously follow the method described herein are not intended for public use and dissemination and shall not be represented as a J1321-Type II test result."[14] It is used by the SmartWay program to assess vehicle and component performance.

SAE J1526 Type III recommended practice remains a work in progress. It is "a standard test procedure for comparing the fuel economy of components or systems" that can be switched from one vehicle to another in a short period of time" and "is ideally suited for comparing the fuel consumption of one vehicle to another and one component of a combination vehicle to the same component in another" (SAE, undated). SAE J1526 Type III is an open-road moderate-distance test with a minimum test distance of 28 mi. It involves two vehicles running the same test without interplay as in SAE J1264 Type I, described below. The object or device being tested is switched from one vehicle to the other on separate runs. Because it is an open-road test, variable control will be more challenging; however, it provides an alternative means of evaluating fuel consumption away from a test track. To improve the accuracy of the results when evaluating devices for the Innovative Technology credits in the Phase I Rule, the minimum route distance for SAE J1526 Type III must be increased from 28 mi to 100 mi and the route "must be representative in terms of grade."[15] Many of the 22 additional provisions from TMC/SAE J1321 Type II could also be applied to SAE J1526 Type III to further improve accuracy. The relative fuel economy of the component or vehicle under test is expressed "as a percentage of fuel saved. This factor is calculated using relative fuel consumption while operating with and without the test component or vehicle under evaluation" (SAE, undated).

SAE J1264 Type I was revised in 2011 and is a recommended practice "providing minimum requirements for testing components or systems of the type which can be switched from one truck to another with relative ease—that is, aerodynamic devices, clutch fans, tires, and the like. The test utilizes in-service fleet vehicles, operated over representative routes."[16] The tests are conducted with two vehicles in simultaneous operation within close proximity of each other: 200-250 yd (180-230 m). Halfway along the test leg of at least 50 mi, the trailing vehicle passes the lead vehicle and remains in the lead position for the remainder of the test leg. The relative fuel effectiveness of the component or system under test is determined as a percentage improvement factor.

The Phase I Rule provides for validation of the drag coefficient (C_d) through coast-down testing. Coast-down tests are usually conducted in accordance with SAE J1263, "Road Load Measurement and Dynamometer Simulation Using Coastdown Techniques." This method was developed for light vehicle testing in relation to dynamometer simulation and to also serve heavy vehicle requirements. An additional recommended practice for coast-down tests is SAE J2263, "Road Load Measurement Using Onboard Anemometry and Coastdown Techniques." The final result of SAE J2263 is a model of road load force (as a function of speed) during operation on a dry, level road under reference conditions.

The precision of the results from these recommended practices is highly dependent on test controls, protocols, and environmental factors. A significant effort is required to ensure that procedures are rigorously followed and that the external factors that can influence fuel consumption are tightly controlled. The fidelity of test results from this J1263 coast-down procedure to results from the powered J1321 (or other) track test could not be established. This is believed to be the critical issue in test procedure selection for trailer regulation.

Wind tunnel testing at reduced scale and full scale is another well-developed method of aerodynamic performance evaluation. SAE J1252 is "Recommended Practice Wind Tunnel Test Procedure for Trucks and Buses." The scope encompasses the full range of full-scale MHDVs represented as either full-scale or reduced-scale wind tunnel models. The document provides guidance for wind tunnel testing to support current vehicle characterization, vehicle development, vehicle concept development, and vehicle component development.

[14] SAE. "Fuel Consumption Test Procedure—Type II." Available at http://standards.sae.org/j1321_201202/. Accessed December 23, 2013.

[15] 40 CFR part 1037.601.

[16] Society of Automotive Engineers, undated. "Joint RCCC/SAE Fuel Consumption Test Procedure (Short Term In-Service Vehicle) Type 1 (STABILIZED May 2011)." Warrendale, Pa.: SAE.

SAE J2971, "Recommended Practice Truck and Bus Aerodynamic Device Terminology" was issued in April 2013. It provides a standard naming convention for aerodynamic devices and technologies used to control aerodynamic forces on trucks, including trailers.

Finally, other promising methods include CFD and the constant-speed torque test. It is generally accepted by experts that constant-speed torque tests require standardization. CFD is proving very useful for product development, but there are practical limitations to its use as a compliance tool for whole vehicle evaluation; however, this is likely to change with time as CFD continues to improve. The recommended practice providing guidance for the use of CFD in evaluating commercial vehicle aerodynamic performance is SAE J2966, "Guidelines for Aerodynamic Assessment of Medium and Heavy Commercial Ground Vehicles Using Computational Fluid Dynamics."

CURRENT USE OF AERODYNAMIC DEVICES AND LOW-ROLLING-RESISTANCE TIRES

Tractors

New tractor specifications throughout the industry have been substantially influenced both by the SmartWay program and the CARB regulation. The committee prepared a questionnaire for several of the largest tractor manufacturers in an effort to quantify this result. The tractor manufacturers that were contacted accounted for about two-thirds of industry sales in the period noted in Table 6-4.

Among those manufacturers surveyed by the committee, nearly 60 percent of tractors sold (first two rows under Share of Sales) are fully equipped sleepers whose fuel consumption already benefits from SmartWay specification. (The manufacturers not contacted, which account for roughly one-third of sales, could have been either more or less likely to produce SmartWay-certified equipment.) This percentage will increase beginning in 2014, as will the performance of other new tractors, since the federal regulations affecting tractors will have become effective.

TABLE 6-4 Industry Sales Penetration of SmartWay Tractors and Components, circa mid-2012 to mid-2013

Tractor Type (Classes 7 and 8)	Share of Sales (%)	Share of S/W[a] Verified Tires (%)	
		S/W Duals	S/W WBSTs
SmartWay fully compliant [b]	38	83	17
Aero sleeper [c]	23	67	17
Day cab w/roof fairing and/or other aero	26	49	5
Day cab, no added fairings	8	49	6
Classic and vocational	4	23	1

NOTE: Percentages are manufacturer sales-weighted.
[a] S/W, SmartWay.
[b] May apply a S/W label.
[c] Short one or more S/W components.
SOURCE: Responses to committee's questionnaire for tractor manufacturers in October 2013.

It is significant that another 26 percent of tractors sold, day cabs, are also aero equipped; half of those use low-rolling-resistance (LRR) tires. Most tractor manufacturers supply well aeroengineered fairings for day cab tractors. It may be possible to separate the use of these tractor types into higher and lower speed applications. In the case of the higher average speed applications, requiring "full" aero-treatment could further reduce fuel consumption. There are certain trailers that are best served by tractors without aeroroof fairings (often called low-roof tractors). Among these trailers are flatbeds and many tankers. Finally, some day cab tractors and some sleeper cab tractors are not single purposed but are used in mixed "utility" haulage. This means pulling a van trailer at times and non-vans at other times; as well, either of those activities leads to performance at a mix of average speeds as duty cycle differences and congestion dictate. Regulators and even carriers need to consider if all these applications can benefit from high aero content.

Further, smart speed recorders might serve as a tool for setting a more stringent aero requirement. (Here the committee imagines a speed recorder computing the cumulative product of tractor moving time and speed-cubed that could be periodically evaluated. Such a time-averaged speed-cubed level would represent the added value of a high-aero configuration.)

A final observation from Table 6-4 is that a full 77 percent of all tractors were equipped with LRR tires,[17] reflecting the carrier's perception of their good value. Probably some of the non-LRRs legitimately require special tire operational features not achievable within an LRR specification.

Aerodynamic Devices on Van Trailers

The committee oversaw the collection of information on the use of aerodynamic devices on trailers by observing nearly 5,000 tractor/trailers operating on interstate highways in seven parts of the country. Persons with knowledge of trucking were provided photographs of the trucks, trailers, and aerodynamic devices of interest and instruction on how to conduct the informal surveys. Observations were made from the side of a highway where traffic could be clearly observed in one or both directions. The results were recorded on the individual forms. The observations were made at the following locations:

[17] Found by multiplying the values in the column Share of Sales by the percentages thereof having duals and WBSTs and summing.

- California, I5, Sacramento, northbound and southbound;
- California, I10, 90 mi east of Los Angles, westbound;
- California, I15, 87 mi east of Los Angeles, southbound;
- California, I80, 93 mi east of Sacramento, westbound;
- Arizona, I10, 10 mi west of Phoenix, eastbound and westbound;
- Oregon, I84, 5 mi east of Portland, eastbound and westbound;
- Texas, I35, 30 mi north of San Antonio, northbound and southbound;
- Michigan, I94, 26 mi west of Ann Arbor, eastbound and westbound;
- Pennsylvania, I81, 29 mi, south of Harrisburg, northbound and southbound; and
- Maryland, I95, 25 mi north of Washington, D.C., northbound and southbound.

The objective was to gain an understanding of the extent to which aerodynamic devices are being used and insight into how much the CARB regulation is influencing utilization of the devices compared to the sole economic motivation of saving fuel. The informal surveys were limited to 53+ ft dry and refrigerated van trailers being pulled by sleeper cab tractors, since this is a common tractor-trailer combination subject to the CARB rule. The exercise did not assess the use of aerodynamic tractors or LRR tires; in any case the latter are difficult to observe while trucks are in operation. Finally, a preliminary assessment did not find use of aerodynamic fairings on the front end of 53+ ft trailers, so such types of devices were not included in the informal survey.

The first major compliance date for trailers subject to the CARB regulation occurred on January 1, 2013. CARB provided an analysis showing that 78 percent of the 53+ ft van trailers should be in compliance; the remainder are allowed a longer phase-in period before compliance is required.[18] It is thus not surprising that observations made in California and Arizona showed a greater proportion of trailers with aerodynamic devices than did those observations made in Oregon, Texas, Michigan, Pennsylvania, and Maryland (Table 6B-1, found in Annex 6B). Side skirts were overwhelmingly the predominant aerodynamic device strategy. Other strategies (underbody fairings and rear fairings) were observed in just a few instances.

Market for Trailer Aerodynamic Devices

Conversations with a selection of major van trailer manufacturers, complemented by the results of the committee's questionnaire (see Annex 6A), provide information on equipping trailers with SmartWay-verified devices. Results summarized in Table 6-5 indicate that 40 percent of new van trailers sold from the four companies surveyed are equipped with side skirts. Several trailer manufacturers produce their own skirts, but they also install third-party skirts upon request. Trailer manufacturers indicate a high customer interest in side-skirt-equipped pup trailers (dual, 28 ft trailers). Several manufacturers reported a fuel use reduction of 7 to 9 percent for pups with side skirts (compared to 5 to 7 percent for advanced sides skirts used on 53+ ft van trailers). Other types of aerodynamic devices are only infrequently used on new trailers.

Trailer manufacturers say that their customers request trailers with aerodynamic devices to comply with the CARB regulation as much as to save fuel. Customers with high gross weight operations have requested trailer weight reduction to increase payload rather than to achieve fuel savings.

Trailer manufacturers also expressed concern that the EPA test procedures do not produce consistent results and say that improvements are needed. Notably, the modified SAE J1321 test procedure is claimed to provide precision of ±1 percent. Manufacturers are concerned that this is inadequate given that average device improvements are in the range of less than 1 percent to about 7 percent. The current precision levels are most troubling when the average performance is in the range of less than 1 to 2 percent. In addition, poor precision may cause a validation result to jump from one bin to another, depending on how the binning is designed. Further, binning ranges could be as narrow as the agencies (EPA and NHTSA) believe is prompted by validation measurement precision and could approach integer values—for example, bins of 4, 5, 6, and so on.

Skirt manufacturers report a growing share of sales to trailer manufacturers (60 percent) rather than retrofitters. Feedback from carriers indicates there is a good ROI from their use.[19] Anecdotal information gained during the observations of on-road trailers indicates a few skirts badly damaged or missing from one side. The skirt manufacturers report no safety concerns (such as side skirts falling off) and little maintenance needed. One skirt manufacturer reported 5-7 percent fuel savings for skirts on container chassis based on fleet tests, although logistical chassis management such as vertical storage at shipyards, horizontal stacking for transport, and variable-length chassis may be problematic.[20] One report on fuel savings from using side skirts on a flatbed trailer said there was a 4 percent reduction in fuel consumption,[21] but loaded product size and shape were unspecified. Flatbeds and container chassis each account for about 7 percent of new trailer sales (ICCT, 2013). An in-depth discussion of benefits

[18] California Air Resources Board (CARB), "Existing and Future Trailer Regulations," personal communication to Tom Cackette and Chuck Salter, committee members, July 30, 2013.

[19] Sean Graham, Freight Wing, Inc., personal communication to Chuck Salter, committee member, August 26, 2013.

[20] Further information is available at http://www.freightwing.com/chassis_fairing.php. Accessed November 18, 2013.

[21] Further information is available at http://freightwing.com/custom_fairing.php. Accessed November 18, 2013.

TABLE 6-5 Characterizing the Experience of Van Trailer Manufacturers Producing SmartWay-Specified Trailers (North America)

		Responses from Companies	S/W Fuel Reduction Performance	Typical Responses
Customer interest in S/W specs	CARB	Yes	-	
	ROI	No	-	
Aero-device volume and prices	Skirts[a]	Total all vans: 40%; all dry vans (range): 25-60%; all reefers (range): 25-40%	≥4% and 5%	$750 and up[d]
	Underbody[b]	Ranges from some to none	5%	$1,000 and up[d]
	Tails[c]	Ranges from some to none	≥1% and 5%	$800 and up[d]
	Gap	None	≥1%	$650 [d]
Aero other than 53 ft vans?		28' vans; 45' vans; and/or tankers	None	Average. Twin 28′ pups performance 7-9%
Customers seek weight reduction?		Yes	-	Yes, for high gross weight operations, not fuel consumption
Collaboration w/tractor OEMs?		Yes, on S/T projects	-	
S/W TP commentary		Level playing field needed; need to specify a standard test tractor; need track round-robin to validate precision	-	
LRR tires usage		LRR duals are offered standard by manufacturers on some trailers (as many as 75%)	≥1.5%	Slightly lower cost than standard but less tread depth
WBST usage		2-10%	≥1.5%	High gross weight; average tandem cost: +$360 w/Fe wheel; +$768 w/Al wheel

NOTE: Al, aluminum; Fe, ferrous; S/W, SmartWay; S/T DOE SuperTruck projects; ROI, return on investment; TP, test procedure; and OEM, original equipment manufacturer.

[a] Skirt data for A,B,C,D combined.

[b]These SmartTruck underbody devices are S/W classified as "advanced trailer end fairings."

[c]S/W categories "trailer boat tails" and "advanced trailer end fairing."

[d] Depending on design, mounting.

SOURCE: Data collected by the committee, August 2013.

and complications associated with other trailer types can be found at NRC (2010, pp. 99-107).

Those manufacturers of trailer tail devices for van trailers that were surveyed by the committee report they sold many fewer such devices than skirts; this was confirmed by the committee's on-road observations. However, one manufacturer reported a 200 percent year-over-year increase in sales. This aerodynamics solution may become the next low-hanging fruit to harvest since fuel savings from these devices is approximately the same as that from side skirts. Fuel savings from trailers equipped with both side skirts and a trailer tail device were reported as 9-10 percent.[22] The tail manufacturer producing a tail-fold-flat feature when the rear doors open claims there are no systematic issues once drivers are properly trained on tail device operation. The fold-flat tail autodeploys when doors are closed by the driver. Figures 6-7, 6-8, and 6-9 show examples of aerodynamic devices.

CARB provided information on the industry that provides aerodynamic devices for trailers. The data in Table 6-6 show that over the past 5 years both the number of companies making aerodynamic devices and the number of devices have increased more than sixfold. Cumulative nationwide sales exceed 400,000 devices, over 90 percent of which are side skirts. Notably, the installed price for trailer side skirts has decreased from $2,800 to less than $1,000, with an ROI break-even point for the truck operator due to fuel savings of less than 11 months.

[22] Jeff Grossman, ATDynamics, personal communication to Chuck Salter, committee member, August 28, 2013.

FIGURE 6-7 Wabash trailer skirt. Available at http://wabash-composites.com/compositeshome/products/trailer-side-skirts/duraplate-aeroskirt-for-truckload-applications. Accessed November 14, 2013.

TABLE 6-6 Trailer Aerodynamic Device Industry

Metric	January 2008	April 2012	June 2013
Companies making SmartWay devices	5	21	33
Number of device types available	11	59	76
Devices sold[a]	>2,200	>180,000	>400,000
Installed cost, average	$2,800	$1,250	<$1,000
Break-even period (months)[b]	30	11	<11

[a] Not all manufacturers contributed data.

[b] Assumes 2.5 trailers per tractor and 105,000 tractor highway miles per year.

SOURCE: California Air Resources Board, "Existing and Future Trailer Regulations," personal communication to Tom Cackette and Chuck Salter, committee members, July 30, 2013.

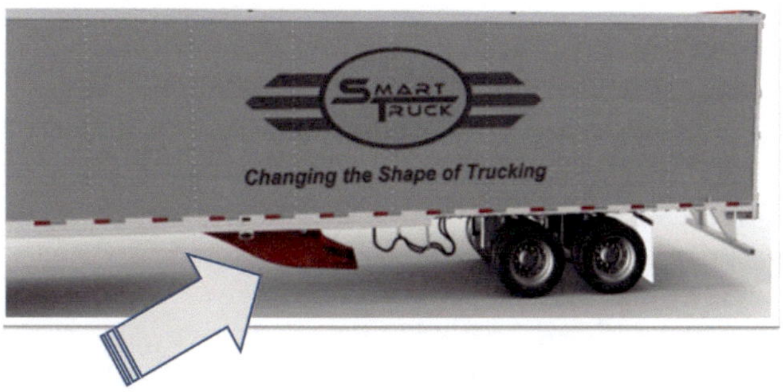

FIGURE 6-8 SmartTruck undertray: UT-1. Available at http://smarttrucksystems.com/undertray-UT1Base.php. Accessed November 14, 2013.

FIGURE 6-9 ATDynamics trailer tail. Available at http://www.atdynamics.com/trailertail.htm. Accessed November 14, 2013.

SmartWay and the CARB regulation apply to the most common tractor and trailer types that are likely to accumulate high mileage at mainly highway speeds, thus taking maximum advantage of aerodynamic improvements. SmartWay verification is limited to high-roof sleeper-cab tractors and 53+ ft dry van trailers, while the CARB regulation also applies to refrigerated 53+ ft van trailers. CARB's requirement to use LRR tires extends to day cabs.

Table 6-7 gives the percentage of trailers that are subject to the SmartWay voluntary program and CARB's regulation of those that are trailers 41 ft and longer and dry van trailers 24 ft and longer, based on new sales in 2012 (ICCT, 2013). About 60 percent of the group of trailers that could benefit from aerodynamic devices and better tires are impacted by these government programs. The other 40 percent of trailers that could benefit from aerodynamic devices such as side skirts and LRR tires include the shorter dry van trailers often used in dual trailer applications, flatbeds, and container chassis (both sea and domestic). The data in the table show that the share of all trailers that could benefit from the use of aerodynamic devices could be as high as 80 percent. Side skirt manufacturers have developed and are marketing skirts for non-van trailers, as discussed previously. Further, smart speed recorders, described at the end of the section "Tractors," might serve to establish if certain of these non-van combination trucks actually operate in a relatively high aerodynamic requirement (one with higher average speeds and annual vehicle miles traveled).

Barriers to Increased Use of Trailer Aerodynamic Devices

Although the fuel savings of pulling a high-use 53+ ft van trailer equipped with side skirts are only about 5 percent at 65 mph, this translates to an annual fuel savings of over $3,500

TABLE 6-7 Trailers Types Not Affected by Government Voluntary and Regulatory Programs

Trailer Category	Length (ft)	Included in[a] SmartWay	CARB	Cumulative Share of Trailer Sales (%)[b,c]
Dry van trailer	≥53	✓	✓	44
Above plus refrigerated van trailers	≥53		✓	47
Above plus midlength van trailers	≥41			54
Above plus short dry van trailers	≥24			66
Above plus flatbed, container, tank trailers	≥41			80

[a] The data shown here include trailers regardless of what type of tractor pulls them.
[b] Of all 41+ foot trailers and 24-40 foot dry van trailers.
[c] New trailer sales in 2012.
SOURCE: ICCT, 2013.

(100,000 mi/year, 5.5 mpg,, and $4/gal). Even if the truck owner has 2.5 trailers per tractor, as is typical, the payback in fuel savings exceeds the cost of installing the side skirts in about 1 year. (A more conservative set of assumptions such as 3 percent fuel savings at 55 mph and 3 trailers per tractor would result in a 16 month payback period, which is still favorable.) So why aren't all trailers, especially new trailers, equipped with fuel-saving devices?

One reason is that the trailer owner may not own the tractor that pulls its trailer. In this situation, the capital cost of trailer modification is borne by the trailer owner, but the fuel savings benefit the tractor owner. Until shipping rates begin to recognize and differentiate the fuel consumption of pulling a trailer with aerodynamic devices and the same consumption without such devices, there is no incentive for a trailer owner who does not own a tractor to purchase aerodynamic-equipped trailers or add aerodynamic devices to existing trailers.

A recent study based on a survey of truck fleet owners and operators and other stakeholders in the supply chain shed additional light on this question. Respondents to the survey identified five main barriers to increased use of more efficient trailers (NACFE and Cascade Sierra Solutions, 2013). These include

- Lack of credible information on fuel savings;
- Uncertainty regarding payback period (capital cost vs. fuel savings);
- Lack of access to capital;
- Questions regarding the reliability of the fuel-saving technologies; and
- No fuel-saving technologies available from preferred trailer or component suppliers.

Lack of credible information was the overarching barrier, which affects the rest of the perceived barriers and in particular the payback period. Small-scale operators were more concerned about the availability of capital than were big fleets operators, as would be expected. The end users were outspoken about the unavailability of product from preferred suppliers, whereas truck and trailer manufacturers did not see this as a significant issue.

Forward trailer bogie positions require shortening skirt length, which reduces aerodynamic performance. This results when trailers are required to meet varying state regulations related to the "bridge formula."

TIRES

The tractor manufacturers surveyed by the committee (Table 6-4) report that over 60 percent of new tractors are sleeper cabs and nearly two-thirds of those are SmartWay-labeled, thus equipped as required with SmartWay-verified LRR tires. Of the remaining one-third not S/W labeled, 84 percent are still ordered with verified LRR tires. This shows that carriers ordering new tractors are aware of and understand the fuel savings from using LRR tires.

The prevalence of WBSTs on sleeper cabs is about 17 percent. There are two incentives for ordering these more efficient tires: further fuel consumption reduction (1-1.5 percent) and weight reduction. Weight reduction due to use of lighter wide-base wheels (aluminum) and tires is especially beneficial in refrigerated ("reefer") trailer applications where high gross weights are typical.

The committee's collection of information from trailer manufacturers (Table 6-5) indicates that LRR tires are standard equipment on a high percentage of new van trailers. Use of WBSTs on new van trailers is in the range of 2 to 10 percent. Trailer manufacturers report that weight reduction is the principal incentive for WBSTs on trailers.

Replacement tire sales in the total commercial MHDV industry outpace those for original equipment application by a factor of three. Sales of replacement tires in 2012 were 15.8 million, while tractor and trailer OEMs purchased 5.1 million (RMA, 2013). Given the great popularity of LRR tires with OEMs, NHTSA should investigate the rolling resistance characteristics of replacement tires sold for these classes of trucks. If a substantial portion of higher rolling resistance replacement tires are in the current sales mix, and if LRR tires are not specified for the operational needs of the legacy fleet, additional fuel savings would accrue if replacement tires had lower rolling resistance.

The committee also sought input from tire manufacturers on the market for LRR truck tires. A questionnaire was developed focusing on tires used on Class 7 and Class 8 combination trucks. The intent was to update the committee's knowledge of the tire industry's contribution and ongoing needs in order to illustrate their valuable role. The industry uptake of WBSTs has been relatively cautious. The committee also sought to better understand the barriers to more widespread adoption of these tires, which typically provide lower rolling resistance than standard width tires.

The Rubber Manufacturers Association (RMA) provided information on conventional tires and WBSTs for both tractors and trailers (Table 6-8), as requested.

For conventional (not wide-base) tires shipped for use on new equipment, the RMA data show about a third lower use of SmartWay LRR tires on new tractors than reported by the tractor OEMs. In any event, the use of LRR tires on new tractors is substantial and is expected to increase significantly as the Phase I Rule requiring GHG emission reductions from new tractors begins. Likewise, many trailer manufacturers report use of SmartWay LRR tires on 75 percent or more of their trailer production, which comports with the RMA data.

The share of SmartWay-verified conventional replacement tires is similar to the share used on new equipment—42 percent of replacement shipments are SmartWay—with greater acceptance on trailers than tractors, according to the RMA data. The committee recognizes there are other operational needs (e.g., traction, scrub resistance, tread life, retreadability) that play a critical role in a carrier achieving lowest life-cycle cost for its tire environment.

WBSTs account for about 3 percent of new tire shipments, for both OEM equipment use and replacements, as shown in Table 6-9. This is lower than reported by OEM tractor manufacturers (approximately 10 percent, see Table 6-4) but within the range of use reported by trailer manufacturers (2 to 10 percent; see Table 6-5). Considering the reported fuel saving of 3 percent or more from replacing conventional narrow tires with wide-base tires (turning an 18-wheeler into a 10-wheeler), the adoption of WBSTs has been slow.

TABLE 6-8 Shipments of New Conventional Narrow Profile Tires for Tractors and Trailers, October 2012 to September 2013 (percent of shipments)

	Application of Conventional Narrow Tires That Are SmartWay-LRR-Verified[a]		
	New Equipment	Replacement Tires	New + Replacement
Tractor	40	37	38
Trailer	79	70	74
Total tractor + trailer	52	42	45

[a]Represents 97 percent of total tire shipments.
SOURCE: Tracey Norberg, Rubber Manufacturers Association.

TABLE 6-9 Shipments of Conventional (Narrow) and Wide-Base Single Truck Tires, October 2012 to September 2013 (percent of shipments)

Application	Conventional Tire	Wide-Base Tire
Tractor (new OE)	97	3
Tractor (replacement tire)	98	2
Trailer (new OE)	96	4
Trailer (replacement tire)	93	7
Tractor + trailer (new OE)	97	3
Tractor + trailer (replacement)	97	3

SOURCE: Tracey Norberg, Rubber Manufacturers Association.

The tire and carrier industry identified the following as issues that may be impeding the greater use of LRR truck tires:

- Because carriers drive tire selection more than do tractor and trailer manufacturers, their highly varying experiences are what most influences purchases.
- The SmartWay program drives industry to seek reduced fuel consumption. However, the current comparative performance data available to carriers may not be precise enough to motivate the purchase of LRR tires.
- Although the ISO28580 rolling resistance test procedure calls for the correlation of measurement results, the industry lacks a master equipment correlation lab that would help assure consistent C_{rr} values are achieved across the industry.
- WBSTs are designated in the United States by the metric sizes 445/50R22.5 and 455/55R22.5. These sizes need to be specified so as to differentiate them from early "super singles," which performed poorly. Super-single issues are discussed in NRC (2010, p. 113).
- At least 50 percent of long-haul tires are retreads. SmartWay retread performance levels were specified by EPA in June 2012. Yet retread C_{rr} performance hurdles (i.e., the maximum value allowed by EPA SmartWay specifications) are higher by 9 to 17 percent than new tire hurdles. And manufacturing and audit controls are believed more lax than for original equipment tires. This represents another opportunity for reducing fuel consumption.

Some barriers to WBST adoption were also identified. Those believed most significant are the following:

- Inability to run flat. Automatic tire inflation systems can significantly reduce this potential hazard. See the next section, "Tire Pressure Systems." (Note that operating on a flat tire is not allowed by FMCSR 393.75.)

- Life-cycle cost remains unsure if the various components such as tread life or downtime due to flats, as well as the comparability of the C_{rr}s themselves, are not correctly estimated.
- Tread life, which—as noted elsewhere—may be as much as a third shorter than that of dual tires.
- The number of allowable retreads is usually one, compared to two and sometimes three for duals.

Overall, about half of SmartWay-verified LRR tire usage is for new equipment and 42 percent for replacement tires. Greater use of LRR tires on new tractors could occur as a result of the Phase I Rule. It is uncertain if greater use of LRR tires on new trailers and for replacement tires will occur in the absence of requirements or incentives.

Tire Pressure Systems

There are two primary types of tire pressure systems: tire pressure monitoring systems (TPMS), which use sensors in various configurations and locations to sense and communicate tire pressure, and automatic tire inflation systems (ATIS), which inflate tires when pressure is low and, in some cases, can deflate tires to correct for pressure rises when the temperature increases. One tire manufacturer remarked that tire pressure systems may be nearly as significant for GHG reduction as lower C_{rr}. One trailer manufacturer reported that 60 percent of its van trailers had been equipped with these systems over the most recent 12 months of production. Likewise, one tire manufacturer reported 40 percent of its trailer tires were equipped with TPS sensors.

These industry reports are corroborated by a U.S. Department of Transportation (DOT) study (2008) on the effectiveness of TPMS. These are some of its findings:

- Improper tire inflation leads to accelerated tire wear (which subsequently leads to compromised braking, poor handling, and reduced stability); increased fuel consumption; greater propensity for catastrophic tire failures (blowouts); more dangerous roadside debris; and an increased number of road calls to repair deflated tires.
- Approximately 7 percent of all tires are underinflated by 20 psi or more. Only 44 percent of all tires are within 5 psi of their target pressure.
- Tire-related costs are the single largest maintenance cost item for commercial vehicle fleet operators.
- Improper tire inflation reduces fuel economy by about 0.6 percent.

Another DOT study (2007) included these important observations:

- Tire pressure monitoring and inflation systems greatly simplify the task of checking and maintaining tire pressure.
- There is significant diversity in the design and technological approach of the marketplace's offering of tire inflation and monitoring systems.
- Commercial vehicle tire inflation and condition directly link to stopping distance and handling and thus to overall safety. Properly maintained and performing tires aid drivers in preventing and mitigating crash situations.

There are 18 or more manufacturers of various types of commercial vehicle tire pressure monitors in North America. Monitor systems are generally characterized by the location of their sensor mounting: on the valve stem (currently the most prevalent in company offerings), on the wheel, or in the tire. Inflation systems are characterized by the nature of the air supply and variable or constant inflation, with the latter the most prevalent in company offerings. Variable inflation pressure is typically utilized to facilitate increased traction, particularly in off-road situations or certain reduced-speed applications. At least a dozen companies offer these products. Both handheld and in-cab readouts are offered.

The North American Council for Freight Efficiency (NACFE) published a report on these systems in August 2013, in which it concluded there are three benefits to carriers for the introduction of such systems, including the decrease in roadside breakdowns due to blowouts caused by underinflation, longer life for the tires, and improved fuel efficiency. These benefits depend on the effectiveness of a fleet's manual tire pressure maintenance system before a device is installed, but the investment in such systems for trailers was recouped in about 8 months. Finally, NACFE estimates that 40 percent of new trailers are being manufactured with TPS, with ATIS outnumbering TPMS by about 3 to 1.

There is a huge body of experience with TPMS in the U.S. light-duty vehicle category. Beginning on October 5, 2005 (and phased in through September 1, 2007), such systems (TPMS) have been required on all four-wheel vehicles up to 10,000 lb GVWR (DOT, 2007).

Following a meeting in the spring of 2011 of ATA's Technology and Maintenance Council, an article appeared in TireBusiness.com entitled "Time for feds to act on truck TPMS."[23] It noted that fleets were continuing to adopt TPMS and ATIS on both tractors and trailers, in acknowledgment that they increase the life of tires and improve fuel economy (test fleets reported a 1.4 percent increase). These improvements came in addition to the safety enhancements reported by the aforementioned DOT study (2007).

The industry article encouraged the FMCSA to soon issue a recommendation for the use of TPMS on commercial

[23] TireBusiness.com, Crain Communication, Inc., "Time for feds to act on truck TPMS," Detroit, Mich., March 28, 2011.

vehicles. That would give fleets direction as to which type of products they should or must use to better monitor and maintain their tires.

The author concluded that by integrating tire monitoring and inflation systems with telematics[24] systems, fleets will greatly improve their tire maintenance, fuel economy, and safety and will reduce their tire costs-per-mile and in-route breakdowns.

A recent response from NHTSA is its solicitation of input on truck tire maintenance practices to help determine the impact of TPS on commercial vehicle fuel economy. This information solicitation is to support its study on feasible fuel-economy standards for medium and heavy-duty trucks for MY2019 and beyond. [25]

FINDINGS AND RECOMMENDATIONS

Trailers

Finding: When a trailer is not owned by the tractor owner-operator (who pays for fuel), there is no incentive for the trailer owner to purchase fuel-saving devices.

Finding: In a survey of trailer manufacturers responsible for two-thirds of industry sales, it was found that only 40 percent of new van trailers come equipped with fuel-saving aerodynamic devices such as side skirts, which suggests that fuel saving is not a dominant consideration in purchasing a new van trailer.

Finding: Only a few van trailer manufacturers promote use of aerodynamic-device-equipped trailers on their websites; others will install devices if requested by the customer, who chooses from an option list.

Finding: The benefits and favorable return on investment that result from more efficient van trailers have been demonstrated by testing and fleet feedback. Use of trailer aerodynamic devices on van trailers, in particular side skirts, provides a full return on investment through fuel savings in about 1 year, on average. Yet the majority of both new and in-use van trailers currently do not use these fuel-saving devices.

Finding: A California regulation requires operators of van trailers to use aerodynamic devices to reduce the energy required to pull them. Observations made in California and Arizona showed a greater proportion of trailers with aerodynamic devices than did those observations made in Oregon, Texas, Michigan, Pennsylvania, and Maryland. Side skirts were overwhelmingly the predominant aerodynamic devices strategy. Other strategies (underbody fairings and rear fairings) were observed in relatively few instances.

Finding: Trailer manufacturers report that compliance with California's regulation is of greater interest than fuel savings when decisions are made on new van trailer purchases. This suggests it is doubtful that the U.S. fleet's use of fuel-efficient trailers will become universal in the absence of a regulation or other strong incentive.

Recommendation 6.1: NHTSA, in coordination with EPA, should adopt a regulation requiring that all new, 53 ft and longer dry van and refrigerated van trailers meet performance standards that will reduce their fuel consumption and CO_2 emissions. The lead time to implement this regulation should be evaluated independently from lead time requirements applicable to the next set of standards for new engines and tractors, because less time is needed to perform compliance testing and install aerodynamic devices on new trailers. The agencies should also collect real-world data on fleet use of aerodynamic trailers to help inform the regulation.

Finding: The current SmartWay program and CARB regulation address only the most commonly used trailer, the 53 ft or longer van trailer, which, among those manufacturers surveyed by the committee, accounts for about 60 percent of the trailers that could benefit from the use of aerodynamic devices. Use of aerodynamic devices on other types of trailers, such as container/chassis and shorter vans, including dual trailers ("pups"), could provide additional fuel savings of 4 to 9 percent per tractor-trailer, according to industry estimates. Fuel savings from the use of side skirts have also been demonstrated on flatbed trailers. The cost-effectiveness of using aerodynamic devices on these additional categories of trailers depends on their annual mileage and average speed, among other considerations such as access to the trailer underbody, and needs further assessment and quantification.

Recommendation 6.2: NHTSA, in coordination with EPA, should determine whether it would be practical and cost-effective to include along with the regulation of van trailers the regulation of other types of trailers such as pups, flatbeds, and container carriers, as doing so could substantially increase overall fuel savings.

Finding: Both trailer and aerodevice manufacturers report that based on replicate tests and testing across different facilities, fuel consumption results determined by the SAE J1321 test procedure lack the necessary precision for accurately assessing the small incremental improvements provided by aerodynamic devices. Depending on the device evaluated, the procedure-specified precision range can be as much as 100 percent of the result.

[24] Denotes the use of devices that incorporate both telecommunications and informatics. See, for example, www.telematics.com.

[25] TireBusiness.com, "NHTSA to Study Mileage Impact of Truck TPMS," December 14, 2012.

Finding: The relative fidelity of test results from the coast-down procedure as opposed to results from a powered on-track test is not known. Fidelity is believed to be the critical parameter in test procedure selection for trailer regulation.

Finding: Aerodynamic practitioners recognize that both wind tunnel testing and computational fluid dynamics (CFD) simulation can provide a good basis for development of aerodynamic surfaces and devices. Both wind tunnel and CFD methods can reduce the cost of development, may avert the building of multiple full-size prototypes, and can shorten development time for the final product.

Finding: The committee's discussions with trailer manufacturers indicate that not all end users are confident of the fuel savings that will be realized from using aerodynamic devices on trailers. Improvement of test precision and repeatability may help address this concern.

Recommendation 6.3: NHTSA should evaluate the relative fidelities of the coast-down procedure and candidate powered procedures to define an optimum prescribed full-vehicle test procedure and process and should validate the improved procedure against real-world vehicle testing. Further, the Agencies should assess if adding yaw loads to the validation process provides significantly increased value to the C_d result. In addition, the Agencies should disseminate to end users updated test data and fuel savings of efficient trailers, aerodynamic devices, and tires, especially to those not participating in the SmartWay program. This should increase end user confidence in fuel savings and device reliability.

Tractors

Finding: Among those manufacturers surveyed by the committee, nearly 60 percent of tractors sold are fully equipped sleepers whose fuel consumption already benefits from SmartWay specification. Because federal regulations that mandate improvements in tractor efficiency begin in 2014, the fraction of efficient tractors sold will likely quickly increase.

Tractors and Trailers

Finding: Tire pressure monitoring systems (TPMS) and automatic tire inflation systems (ATIS)—collectively, "tire pressure systems," or TPS—are being increasingly accepted by fleets for their operational benefits (fewer flats, reduced fuel consumption, and longer tire life) and considerable safety benefits (uncompromised vehicle stability and short stopping distances).

Finding: The commercial vehicle industry has no standards for any of the following: TPS designation, minimum performance for various system types, driver displays, and testing procedures for system validations.

Recommendation 6.4: The lack of standardization of TPS for commercial vehicles might be appropriately remedied by professional or industry organizations such as the Society of Automotive Engineers or the Technology and Maintenance Council, or by a collaboration of such organizations. In support of that activity, it would be beneficial for NHTSA to prepare a white paper to clarify the minimum TPS performance needed from a safety perspective.

Tires

Finding: Many new tractors and most new trailers are equipped with low-rolling-resistance tires that meet the SmartWay performance standard, and the share of vehicles so equipped is likely to increase owing to regulatory requirements. However, 70 percent of new tires sold in 2012 for use on tractors and trailers were for replacement of existing tires, and only 42 percent of these were SmartWay verified. There is no assurance that replacement tires will one day be as energy efficient as the original equipment tires they replace.

Recommendation 6.5: NHTSA, in coordination with EPA, should further evaluate and quantify the rolling resistance of new tires, especially those sold as replacements. If additional cost-effective fuel savings can be achieved, NHTSA should adopt a regulation establishing a low-rolling-resistance performance standard for all new tires designed for tractor and trailer use.

Finding: C_{rr} measurement in tires will have to be precise given the relatively modest fuel savings achievable with low-rolling-resistance tires. Further, while the ISO28580 test procedure is given good grades by most in the industry, there does not exist a robust machine cross-correlation for commercial vehicle tires in the United States. Carriers cannot depend on the comparability of C_{rr} measurements from the approximately 60 tire suppliers verified by SmartWay.

Recommendation 6.6: NHTSA, supported by EPA, should expeditiously establish and validate the equipment and process for a tire industry machine alignment laboratory and mandate the use of that laboratory by each tire manufacturer seeking C_{rr} validation for any tires being offered as candidates in the GEM computation process, just as the C_{rr}'s of light-duty-vehicle tires were validated.

REFERENCES

Corbett, J.J., and J.J. Winebrake. 2007. Sustainable movement of goods: Energy and environmental implications of trucks, trains, ships, and planes, *Environmental Management* (November): 8-12.

Cummins, Inc. 2009. Framework for the Regulation of GHGs from Commercial Vehicles.

Department of Transportation (DOT). 2007. Tire Pressure Monitoring and Maintenance Systems Performance Report, FMCSA-PSV-07-001. Washington, D.C.: U.S. Department of Transportation, January.

DOT. 2008. Commercial Vehicle Safety Technologies: Applications for Tire Pressure Monitoring and Management. Paper Number 09-0134. Washington, D.C.: U.S. Department of Transportation.

EPA and NHTSA. 2011. Federal Register, Vol. 76, No.179, pp. 57106 to 57513. September 15.

Federal Highway Administration (FHWA). 2006. *Freight Performance Measure: Travel Time in Freight Significant Corridors.* Washington, D.C.: U.S. Department of Transportation.

International Council on Clean Transportation. 2013. Trailer technologies for increased heavy-duty vehicle efficiency—Technical, market, and policy considerations. White Paper. June.

National Research Council (NRC). 2010. *Technologies and Approaches to Reducing the Fuel Consumption of Medium- and Heavy-Duty Vehicles.* Washington, D.C.: The National Academies Press.

NRC. 2012. *Review of the 21st Century Truck Partnership, Second Report.* Washington, D.C.: The National Academies Press.

North American Council for Freight Efficiency (NACFE) and Cascade Sierra Solutions. 2013. *Barriers to the Increased Adoption of Fuel Efficiency Technologies in the North American On-Road Freight Sector.* Fort Wayne, Ind.: NACFE. July.

NACFE. 2013. Tire Pressure Systems—Confidence Report. Fort Wayne, Ind.: NACFE. August.

Rubber Manufacturers Association (RMA). 2013. 2012 Tire shipments unchanged. Dan Zielinski, Washington, D.C., April 3.

Society of Automotive Engineers. Undated. "Fuel Consumption In-Service Test Procedure Type III." Warrendale, Pa.: SAE International.

TIAX. 2009. *Assessment of Fuel Consumption Technologies for Medium- and Heavy-Duty Vehicles.* Report prepared for the National Academy of Sciences by TIAX LLC, Cupertino, Calif. July 31.

Winebrake, J.J., J.J. Corbett, A. Falzarano, J.S. Hawker, K. Korfmacher, S. Ketha, and S. Zilora. 2008. Assessing energy, environmental, and economic tradeoffs in intermodal freight transportation. *Journal of the Air and Waste Management Association* 58(8): 1004-1013.

ANNEX 6A
QUESTIONS POSED TO VAN TRAILER MANUFACTURERS TO GATHER INFORMATION FOR TABLE 6-5

Briefly describe your product type(s), sales volumes, and typical selling price (OE and retrofit) for each (perhaps you want to report the highest 1 or 2 performing devices in each SmartWay category in which you produce).

Identify sales volumes driven by (1) Customers' Expected ROI, (2) Compliance to California Regulation, (3) Unknown.

For described products, report SmartWay aerodynamic performance (and test procedure used for each: J1321 track, wind tunnel, coast-down, other):

Customer acceptance. Please comment on what you believe are the biggest barriers to acceptance of each product type—customers lacking credible performance information; installed cost (customer uncertainty of information for ROI calculation; poor ROI), cost of performance certification; interference with normal driver functions; existing government regulations; other (specify) (list in approximate descending order). Include customer feedback.

Describe any customer feedback you have received on maintenance or reliability issues. Are they resolved?

What safety concerns or experiences have been reported by customers?

Do you anticipate further performance improvements for any of the described product types in the next 5+ years? (Please be as specific as you can.)

Are you collaborating with tractor OEMs in product development?

Which corporate entity do you think should be responsible for trailer performance certification in the event of a greenhouse gas regulation that includes trailers? Tractor, trailer, or device manufacturer? Please explain your thinking.

Are your products applicable to trailer types other than 53 ft vans? What marketing incentives are absent that currently restrict those developments? Will you share aerodynamic performance results with your products on trailer types other than 53 ft vans?

ANNEX 6B

This annex reports the raw data from the observations of van trailers observed by 12 experts during August 2013 (Annex Table 6B-1). The contributors made these observations working at 10 locations, individually or in teams. The day and time of the observations was at the discretion of these individuals and did not follow a pre-set sampling plan, so the results of this informal survey cannot be generalized. The accuracy of these observations was not verified. The locations were chosen to include five locations in or near California; the major cross-country routes into California (e.g., I-80, I-15) and another route leading into California from Arizona (I-10); and five locations in different parts of the continental United States. The purpose was to explore the hypothesis that van trailers operating in, or destined for, California, which is the only state currently requiring the use of aerodynamic devices, would have a higher observed incidence of aerodynamic trailer fittings than those in other parts of the country.

ANNEX TABLE 6B-1 Observations Made on the Use of Aerodynamic Devices on 53+ ft Dry or Refrigerated Vans Pulled by Sleeper Cab Tractors

Location[a]	No. of Trailers in Sample	Side Skirts	Underbody Fairings	Rear Fairing	Trailer w/ Device[b]
CA, I-5, Sacramento, NB, SB	565	199	11	0	210
CA, I-10, 90 mi E of LA, WB	1,068	371	42	10	414
CA, I-15, 87 mi E of LA, SB	944	392	55	8	450
CA, I-80, 93 mi E of Sacramento, WB	497	206	15	2	216
AZ, I-10, 10 mi W of Phoenix, EB, WB	300	119	10	6	130
OR, I-84, 5 mi E of Portland, EB, WB	100	23	1	1	24
TX, I-35, 30 mi N of San Antonio, NB, SB	215	45	4	3	49
MI, I-94, 26 mi W of Ann Arbor, EB, WB	289	64	2	0	66
PA, I-81, 29 mi S of Harrisburg, NB, SB	662	170	11	0	181
MD, I-95, 25 mi N of Washington DC, NB, SB	300	68	6	0	74

NOTE: See note in text regarding the design of the observations. AZ, Arizona; CA, California; DC, District of Columbia; MD, Maryland; MI, Michigan; PA, Pennsylvania; and OR, Oregon.

[a] WB, SB, and so on indicate the direction on the freeway surveyed. Some surveys were in only one direction.

[b] Trailers using one or more devices. The sum of columns 3, 4, and 5 does not agree with the final column because a small number of trucks use two devices.

Appendixes

A

Committee Biographical Information

ANDREW BROWN, JR. (NAE) is vice president and chief technologist, Delphi Corporation. Dr. Brown came to Delphi from the GM Research and Development Center in Warren, Michigan, where he was director of Strategic Futures. He served as manager of the Saturn Car Facilities from 1985 to 1987. At Saturn, he was on the Site Selection Team and responsible for the conceptual design and engineering of this innovative manufacturing facility. Dr. Brown began his GM career as a project engineer at Manufacturing Development in 1973. He progressed in the engineering field as a senior project engineer, staff development engineer, and manager of R&D for the Manufacturing Staff. During this period, he worked on manufacturing processes and systems with an emphasis on energy systems, productivity improvement, and environmental efficiency. Before joining GM, he supervised process development at Allied Signal Corporation, now Honeywell, Incorporated, in Morristown, New Jersey. He earned a B.S. in chemical engineering from Wayne State University in 1971. He then received an M.B.A. in finance and marketing from Wayne State University in 1975 and an M.S. in mechanical engineering with a focus on energy and environmental engineering from the University of Detroit-Mercy in 1978. He completed the Penn State Executive Management Course in 1979. A registered professional engineer, Dr. Brown earned a doctorate of engineering in September 1992. He is currently or has served on the boards of the following organizations: Society of Automotive Engineers, Inc.; Engineering Society of Detroit; Convergence Education Foundation, National Inventors Hall of Fame, Convergence Transportation Electronics Foundation; National Council of Engineering Examiners; State of Michigan Board of Professional Engineers; and the WSR College of Engineering board of advisors. He is a member of the National Research Council's Board on Energy and Environmental Systems and recently served as a member of the NRC's Committee on Review of the 21st Century Truck Partnership and chaired the Committee on Fuel Economy of Medium- and Heavy-Duty Vehicles. Dr. Brown has been an adjunct professor at Wayne State University, the University of Michigan, and Tsinghua University (Beijing, China).

INES AZEVEDO is the codirector of the Center for Climate and Energy Decision Making at Carnegie Mellon University (CMU) and an associate professor with CMU's Department of Engineering and Public Policy. Dr. Azevedo's research interests lie at the intersection of environmental, technical, and economic issues, such as how to address the challenge of climate change and to move towards a more sustainable energy system. In particular, she has been looking at how energy systems are likely to be shaped in the future, which requires comprehensive knowledge not only of the technologies that can address future energy needs but also of the decision-making process followed by different agents in the economy. She received her B.Sc in environmental engineering from IST University in Portugal, her M.Sc. in engineering policy and management of technology from IST, and her Ph.D. from CMU in engineering and public policy.

RODICA BARANESCU (NAE) is a professor in the College of Engineering, Department of Mechanical and Industrial Engineering, University of Illinois at Chicago. Before that, she was manager of the fuels, lubricants, and engine group of the International Truck and Engine Corporation, at Melrose Park, Illinois. She is an internationally sought after public speaker on technical issues related to mobility technology, environmental control, fuels, and energy. She has extensive expertise in diesel engine technology and was elected to the NAE in 2001 for research leading to effective and environmentally sensitive diesel and alternative-fuel engines and leadership in automotive engineering. She is a fellow of SAE International and was its president in 2000. In 2003 she received the Internal Combustion Engine Award of the American Society of Mechanical Engineering (ASME). Dr. Baranescu received her M.S. and Ph.D. degrees in mechanical engineering in 1961 and 1970, respectively, from the Politehnica University in Bucharest, Romania, where she served as assistant professor (1964-1968), lecturer (1970-1974), and associate professor (1974-1978).

THOMAS CACKETTE retired at the end of 2012 after serving as the chief deputy executive officer of the California Air Resources Board for over 20 years. With the Board since 1982, he managed the Board's motor vehicle emission control program, which develops regulations and other programs to reduce vehicle emissions. He also managed the Board's Monitoring and Laboratory Division, which performs ambient air quality monitoring and develops test methods. Overall, 400 professional and support staff are dedicated to these programs, which are contributing to a steady decline in air pollution in California's major urban areas. Mr. Cackette has been involved in many aspects of air pollution control since 1974. He served as a legislative lobbyist for the ARB for several years, and worked 8 years for the U.S. Environmental Protection Agency Motor Vehicle Emission Laboratory in a variety of technical, management, and policy positions. Prior to that he was involved in rocket engine production, testing, and flight performance analysis at Rocketdyne in Los Angeles, where he gained firsthand knowledge of living in the smoggiest city in the United States. He holds an M.S. in engineering and a B.S. in aeronautics and astronautics. He has published papers for the Society of Automotive Engineers and the Air and Waste Management Association and is a frequent speaker on air quality issues.

NIGEL N. CLARK is the George B. Berry Chair of Engineering in the Statler College of Engineering & Mineral Resources at West Virginia University (WVU) and associate vice president for Academic Strategic Planning at WVU. He holds a Ph.D. in chemical engineering from the University of Natal, South Africa, and previously held assistant and associate professor positions in the Department of Mechanical and Aerospace Engineering at WVU. Dr. Clark's areas of interest include vehicle design, advanced vehicle concepts, alternative fuels, and the measurement and reduction of vehicle emissions. He has also published extensively in the areas of particle science and multiphase systems. He has conducted research for government and industry in the areas of fuel economy and emissions from heavy-duty vehicles, including buses and heavy hybrid drive vehicles, and works with the International Council for Clean Transportation on technology and efficiency review. Dr. Clark has contributed to understanding the influence of vehicle activity and test cycles on fuel use and to relating engine and vehicle dynamometer data. He commenced his career with a National Science Foundation Presidential Young Investigator Award, was recognized as a Benedum Scholar by his institution, and is a fellow of the Society of Automotive Engineers.

RONALD GRAVES is the director of the Sustainable Transportation Program at the Oak Ridge National Laboratory (ORNL), which covers the laboratory's research in vehicle efficiency technologies, fuels, and intelligent transportation systems. He joined ORNL in 1976 after receiving his Ph.D. in mechanical engineering from the University of Tennessee. He has been a member and leader of many projects in transportation fuels and engines since the early 1980s, including work on pathways to higher engine efficiency, alcohol fuels, and the effects of fuel sulfur and fuel composition on combustion and emissions. He currently serves on the technical teams for the U.S. DRIVE Partnership and the 21st Century Truck Partnership. Over 25 years ago, he led the establishment of engine and emissions research at ORNL that continues today. He has participated in working groups of the Coordinating Research Council and is a fellow of the Society of Automotive Engineers. Dr. Graves has a record of over 60 publications and reports that encompass subjects in internal combustion engines, fuels, power systems, and materials. He shares four patents with coworkers.

DAN HANCOCK (NAE) retired from GM in 2011. Since June 2010 he had been GM vice president, Global Strategic Product Alliances. In that newly created position, Mr. Hancock was charged with building strong product alliance relationships and speeding development and implementation of joint ventures for winning vehicles and technologies. His previous appointments included GM Powertrain vice president, global engineering and powertrain, and chief executive officer, Fiat-GM Powertrain, based in Turin, Italy. After joining General Motors in 1968 he held various engineering positions within Allison Transmission Division. In 1983 he became chief engineer for Detroit Diesel in Redford, Michigan. He became technical director, Advanced Powertrain, at the Chevrolet-Pontiac-GM Canada Group in 1987. In 1992, he was appointed chief engineer of the Small Block V8 engine and in 1994 was appointed director, transmission engineering, GM Powertrain. In 1997 he returned to Indianapolis, where he was named president, Allison Transmission Division. Mr. Hancock received a master's degree in mechanical engineering from Massachusetts Institute of Technology in 1973 and a bachelor's degree, also in mechanical engineering, from the General Motors Institute, Michigan, in 1974. He served as chairman of the Society of Automotive Engineers Foundation Board of Trustees from 1998 to 2008. He served as president of FISITA, the International Federation of Automotive Engineering Societies, from 2004 to 2006. He was inducted into the National Academy of Engineering in 2011. He is a recipient of the SAE Medal of Honor, the Great Golden Medal for Service to the Republic of Austria, and the Sagamore of the Wabash recognition from the State of Indiana. He has been elected SAE president for 2014.

W. MICHAEL HANEMANN (NAS) joined the Arizona State University (ASU) Department of Economics and the Center for Environmental Economics and Sustainability Policy in 2011, where he is a Wrigley Chair in Sustainability. He came to ASU from the University of California, Berkeley, where he was a Chancellor's Professor in the Department of Agricultural and Resources Economics and the Goldman School of Public Policy. His research interests include nonmarket valuation, the economics of water and of climate change, environmental policy, adaptive management, and demand modeling for market research. Dr. Hanemann has served on many NRC committees and was elected to the National Academy of Sciences in 2011. He is currently a lead author and a contributing lead author for Working Group III of the IPCC Fifth Assessment Report on Climate Change. Dr. Hanemann received his B.A. from Oxford University in philosophy, politics, and economics, his M.S. from the London School of Economics in development economics, and his M.A. and Ph.D. from Harvard University in public finance and decision theory and economics. He received an honorary Ph.D. from the Swedish University of Agricultural Sciences and the Lifetime Award for Outstanding Achievement from the European Association of Environmental and Resource Economists. He is an inaugural fellow of the Association of Environmental and Resource Economists and a fellow of the American Association of Agricultural Economics.

WINSTON HARRINGTON is senior fellow at Resources for the Future, where his research interests include urban transportation, motor vehicles and air quality, and problems of estimating the costs of environmental policy. He has worked extensively on the economics of enforcing environmental regulations, the health benefits derived from improved air quality, the costs of waterborne disease outbreaks, endangered species policy, federal rulemaking procedures, and the economics of outdoor recreation. Dr. Harrington has written or coauthored five books and numerous book chapters. In October 2000, he won the Vernon Award of the Association of Public Policy Analysis and Management for a paper he coauthored, "On the Accuracy of Regulatory Cost Estimates." He has served as a consultant to U.S. state and federal governments, the World Bank, and the Harvard Institute for International Development and has worked in Lithuania, Mexico, and Poland.

GARY MARCHANT is a Regents' Professor of Law and faculty director of the Center for Law, Science, and Innovation in the College of Law at Arizona State University. He is also a Senior Sustainability Scientist at ASU's Global Institute of Sustainability. Professor Marchant teaches environmental law, science and technology, genetics and the law, and environmental justice. Prior to joining the ASU faculty, he was a partner at the Washington, D.C., office of the law firm Kirkland & Ellis, where his practice focused on environmental and administrative law. He received his B.Sc. and Ph.D. in genetics from the University of British Columbia, his M.P.P. from the Kennedy School of Government of Harvard University, and his J.D. from Harvard Law School.

PAUL MENIG is CEO of Tech-I-M, a consultancy. Previously he was employed by Freightliner, where he was responsible for daily production problems, field problems, custom work orders, and advanced engineering for electrical and electronic items such as engines, transmissions, brakes, and safety devices. Mr. Menig joined Daimler Trucks North America in July 1994 and initially led the development of electronics for the new Freightliner Century Class truck product line. Prior to joining Freightliner, Mr. Menig spent 7 years with Eaton Truck Components, leading a team of as many as 65 people in the development of electronic products for automated mechanical transmissions, brakes, and tire pressure control. These activities included some worldwide responsibility and coordination with engineering in Europe and joint venture development with Japanese companies. Prior to that, Mr. Menig worked for the industrial automation part of Eaton known as Cutler-Hammer. During those 8 years he led teams working on sensors, factory communications, programmable and motion controllers, and vision inspection equipment. Prior to Eaton, Mr. Menig worked 5 years for General Electric in the areas of medical equipment for hospitals, remotely guided military vehicles (smart bombs), and charge-coupled device imagers and signal processors. Mr. Menig graduated from MIT in 1976 with a bachelor's degree in electrical engineering. He participated in the ABC program of General Electric, completing the A and B portions. Master's degree work in electrical engineering was completed with the exception of a thesis at Marquette University. In addition, Mr. Menig has participated in numerous training programs such as total quality management, software development, strategic planning, finance for the nonfinancial manager, ISO 9000, and vehicle dynamics.

DAVID F. MERRION is chairman of Merrion Expert Consulting LLC and the retired executive vice president of engineering for Detroit Diesel Corporation (DDC), a subsidiary of Daimler Trucks North America. His positions at DDC included staff engineer, Emissions and Combustion; staff engineer, Research and Development; chief engineer, Applications; director, Diesel Engineering; general director, Engineering (Engines and Transmissions); and senior vice president, Engineering. Mr. Merrion has extensive expertise in the research, development, and manufacturing of advanced diesel engines, including alternative-fueled engines. He is a Society of Automotive Engineers fellow and a member of the American Society of Mechanical Engineers. In 2009 he received the ASME Honda Medal and in 2012 he received the SAE Powertrain Innovation Award. He served as president of the Engine Manufacturers Association, a member of EPA's Mobile Sources Technical Advisory Committee, a member of the Coordinating Research Council; and a member of the U.S. Alternate Fuels Council. He has served on a number of National Research Council committees, including the Standing Committee to Review the Research Program of the Partnership for a New Generation of Vehicles; the Committee on Review of the 21st Century Truck Partnership, Phase 1 and Phase 2; and the Committee to Assess Fuel Economy Technologies for Medium- and Heavy-Duty Vehicles. He has a bachelor of mechanical engineering from General Motors Institute (Kettering University) and holds an M.S. in mechanical engineering from the Massachusetts Institute of Technology.

AMELIA REGAN is a professor of computer science and civil (transportation systems) engineering at the University of California, Irvine. Her research interests include dynamic and stochastic network optimization, parallel and distributed computing, optimal contracting, port operations, logistics systems analysis, freight industry analysis, shipper behavior modeling, freight transportation planning, combinatorial and online auction mechanism and algorithm design, transportation economics, data mining, vehicle-to-vehicle and vehicle-to-roadside communication systems (VANets), network design under uncertain demand, humanitarian cyber-physical systems, and cyberphysical transportation systems. Dr. Regan's research has been supported by the National Science Foundation, the Transportation Research Board, JB Hunt, Transportation Inc., the University of California Transportation Centers, and the CalTrans PATH program. It has been published in more than 120 refereed journal articles and conference proceedings papers in *Transportation Research* (A, B, C and E), *Transportation Science, Operations Research, INFOR, the Transportation Research Record, the Transportation Journal, Transportation, Computers and Industrial Engineering, IEEE Network,* and *IIE Transactions*, among others. She has been at UCI since 1997, where she has had primary faculty appointments in the Departments of Computer Science and Civil and Environmental Engineering and a courtesy appointment in the Paul Merage School of Business (formerly the Graduate School of Management). She was the associate dean for student affairs for the Bren School of Information and Computer Sciences from 2005-2009. Previously she earned M.S. and Ph.D. degrees in Transportation Systems Engineering from the University of Texas, Austin, an M.S. in applied mathematics from the Johns Hopkins University, and a B.A.S. in systems engineering from the University of Pennsylvania. She also worked as a research engineer, software engineer, and operations research analyst for the Association of American Railroads and United Parcel Service prior to joining the Ph.D. program at the University of Texas.

MIKE ROETH is executive director of the North American Council for Freight Efficiency and has worked in the commercial vehicle industry for over 28 years. He is also leading the Trucking Efficiency Operations for the Carbon War Room. Mr. Roeth's specialty is brokering green truck collaborative technologies into the real world at scale. As director, Global Advanced Engineering, for Navistar International, he led the advanced engineering efforts for the Navistar family of vehicle brands: International trucks, Navistar Defense, IC buses, and Workhorse Custom Chassis. These efforts included fuel economy improvement, emissions reduction, driver comfort and efficiency, as well as quality, cost, and performance breakthroughs. He has a B.S. in engineering from the Ohio State University and a master's in organizational leadership from the Indiana Institute of Technology. Mr. Roeth is a 27 year member of the Society of Automotive Engineers; he also served as a board member of the Automotive Industry Action Group and as chairman of the board for the Truck Manufacturers Association. He also has been heavily involved with the 21st Century Truck Partnership, a collaborative effort between industry and the U.S. government and the Hybrid Truck Users Forum of CALSTART.

GARY W. ROGERS is an independent consultant. Previously, he was president, chief executive officer, and sole director, FEV, Inc. His previous positions included director, Power Plant Engineering Services Division, and senior analytical engineer, Failure Analysis Associates, Inc.; design development engineer, Garrett Turbine Engine Company; and Exploration Geophysicist, Shell Oil Company. He has extensive experience in research, design, and development of advanced engine and powertrain systems, including homogeneous and direct-injection gasoline engines, high-speed direct-injection passenger car diesel engines, heavy-duty diesel engines, hybrid vehicle systems, gas turbines, pumps, and compressors. He provides corporate leadership for a multinational research, design, and development organization specializing in engines and energy systems. He is a fellow of the SAE, is an advisor to the U.S. Army Tank Automotive Research Development and Engineering Center (TARDEC) on heavy-fuel engines, and sits on the President's Advisory Board of Clemson University and on the advisory board to the College of Engineering and Computer Science, Oakland University. He is currently a member of the Board on Energy and Environmental Systems for the National Academies and has served as a member of many NRC bodies, including the Committee on Review of DOE's Office of Heavy Vehicle Technologies Program, the Committee on the Effectiveness and Impact of Corporate Average Fuel Economy (CAFE) Standards, the Panel on Benefits of DOE's Light-Duty Hybrid Vehicle R&D Program, the Committee for the Assessment of Technologies for Improving Light Duty Vehicle Fuel Economy, and the Committee to Review the 21st Century Truck Partnership. He holds a B.S. in mechanical engineering from Northern Arizona University and an M.E. in mechanical engineering from the University of Colorado.

CHARLES K. SALTER is retired after 39 years with Mack Trucks, Inc./Volvo PowerTrain NA. His experience covers a wide range of heavy-duty diesel engine engineering and development. His most recent positions included executive director, Engine Development, where he was responsible for all engine/system functions (design and analysis; emissions control/fuel economy; electronics systems and test). This responsibility included the design and production introduction of the world's first fully electronically controlled diesel unit injector engine. He was also executive director, Advanced Engine Engineering, and collaborated with three-site (Sweden, France, United States) advanced heavy-duty diesel engine research projects. He jointly initiated (with Detroit Diesel) and developed, with EPA and industry, a urea infrastructure for 2007 engine production (then delayed to 2010). He participated in industry collaborative research through the DOE Diesel Crosscut Committee, part of the 21st Century Truck Partnership. He was a consultant to Volvo PowerTrain NA on advanced large truck diesel exhaust gas recirculation cooler vibration study/amelioration; on heavy-duty truck hybrid powertrain duty cycle test procedure development for comparative fuel consumption (EPA/industry/Hybrid Truck Users Forum), and a study of regulatory boundaries for the EPA heavy-duty truck and engine nonconformance penalty rule. He has served on two National Research Council committees, including the Committee to Assess Fuel Economy Technologies for Medium- and Heavy-Duty Vehicles and the Committee on Review of the 21st Century Truck Partnership Phase 2. He has been a Society of Automotive Engineers member for 47 years and was a board member of the Engine Manufacturers Association for 25 years, including a term as president. He was a member of Technology and Maintenance Council of the American Trucking Associations. He holds a B.S. in mechanical engineering from Penn State University and an M.S. in mechanical engineering, solid mechanics, from the University of Maryland.

CHRISTINE VUJOVICH retired in 2009 from Cummins, Inc., as its vice president of marketing and environmental policy. During much of her 31 years at Cummins, Mrs. Vujovich served as its environmental policy officer. In the late 1980s, she attended to the heavy-duty engine issues for her company in the reauthorization of the Clean Air Act. She collaborated with other industry representatives to develop industry positions and worked with congressional staff to balance new legislative initiatives with technology practicalities. Those initiatives included NO_x and particulate requirements for heavy-duty engines. Her experiences with the U.S. regulatory process aided her similar work in Europe, China, and India. She oversaw Cummins' participation in the development of several automotive regulations for the commercial vehicle and equipment markets and then oversaw the implementation of these regulations through the technical work inside the company. During her tenure at Cummins, she served terms as chair for the Engine Manufacturers Association and the Mobile Source Technical Review Subcommittee of the Clear Air Advisory Committee. Since retiring from Cummins, Mrs. Vujovich has co-chaired the Health Effects Institute Special Committee on Emerging Technology, whose work culminated in the publication *The Future of Vehicle Fuels and Technologies: Anticipating Health Benefits and Challenges*. She has in addition served as adjunct faculty at the Indiana University School of Public and Environmental Affairs. Mrs. Vujovich has an undergraduate degree in the teaching of the earth sciences and a master's degree in environmental engineering from the University of Illinois. She attended the Yale executive management program.

JOHN WOODROOFFE heads the Transportation Safety Analytics program and is director of the Commercial Vehicle Research and Policy Program at the University of Michigan Transportation Research Institute (UMTRI). He is responsible for the Center for National Truck and Bus Statistics, which conducts nationwide surveys of Trucks Involved in Fatal Accidents (TIFA) and Buses Involved in Fatal Accidents (BIFA), and the Statistical Analysis Group, which performs analytical modeling and conducts research to advance statistical methods for road and vehicle safety analysis. He is an international expert on policy and safety evaluation of large vehicles, including stability and control, accident reconstruction, vehicle productivity, fuel use, and environmental impact. He has participated in many large international technical projects and has been a member of vehicle-related technical expert working groups of the Organisation for Economic Cooperation and Developemnt (OECD), most recently the OECD/JTRC project entitled "Heavy Vehicles: Regulatory, Operational and Productivity Improvements." This Paris-based international task force examined regulatory concepts and future truck technology and sustainable road transport. Prior to joining UMTRI, Mr. Woodrooffe founded the Road Vehicle Research Program at the National Research Council of Canada and developed it into a successful, internationally active heavy truck research laboratory. He was a consultant to Australia's National Road Transport Commission for a unique 3-year performance-based standards development project that produced a new performance-based regulatory system for large vehicle combinations. Mr. Woodrooffe holds master's and bachelor's degrees in mechanical engineering from the University of Ottawa.

MARTIN ZIMMERMAN is the Ford Motor Company Clinical Professor of Business Administration at the Ross School of Business at the University of Michigan. His career has spanned academia, government, and business. He served as chief economist as well as group vice president at Ford Motor Company, where he was responsible for corporate economics, governmental affairs, environmental and safety engineering, and corporate social responsibility. Before joining Ford, he taught at the Sloan School of Management at MIT and at the business school at the University of Michigan. He served on the National Commission on Energy Policy and also served as a senior staff economist on the President's Council of Economic Advisors and as a member of the Panel of Economic Advisors to the Congressional Budget Office. He is presently the vice chair of the Board of the National Bureau of Economic Research. His research is concerned with energy policy, government regulation of business, and economic developments in the automotive industry. Professor Zimmerman earned a Ph.D. in economics from MIT and the A.B. degree from Dartmouth College.

B

Statement of Task

1. The committee will review the NHTSA fuel consumption regulations promulgated on 15 September 2011 (76 Federal Register 57106) and consider the technological, market and regulatory factors that may be of relevance to a revised and updated regulatory regime taking effect for model years 2019-2022. This review will include, but not be limited to, the potential for technological change in commercial vehicles in MY 2019-2022 and the impact it might have on the regulatory regime. Also as part of its review, the committee will explore regulatory options for trailers.
2. The committee will analyze and provide options for improvements to the certification and compliance procedures for medium- and heavy-duty vehicles—including the use of representative test cycles and simulation using various models—such as might be implemented in revised fuel consumption regulations affecting MY 2019-2022.
3. The committee will review an updated analysis of the makeup and characterization of the medium- and heavy-duty truck fleet, including combination tractors, trailers, busses and vocational vehicles. The committee also will review the methodology for providing on-road information on fuel consumption.
4. The committee will examine the barriers to and the potential applications of natural gas in class 2b through class 8 vehicles. The committee will consider how such vehicles could be included in the framework on fuel consumption regulations.
5. The committee will address uncertainties and perform sensitivity analyses for the fuel consumption and cost/benefit estimates, to the extent possible, and provide guidance to NHTSA on improving its uncertainty analyses given the relatively long time frame for these future estimates.

C

Committee Activities

COMMITTEE MEETING, MARCH 20-21, 2013, WASHINGTON, D.C.

NRC Assessment of Technologies for Reducing the Fuel Consumption of Medium- and Heavy-Duty Vehicles: Briefing for the NRC Committee
James Tamm, chief, Fuel Economy Division, National Highway Traffic Safety Administration

Medium and Heavy Duty Fuel Efficiency and GHG Emission Standards: Phase 1 Overview and a Look Ahead to Phase 2
Matthew Spears, center director, Heavy-Duty Diesel Standards, Environmental Protection Agency (EPA)

21st Century Truck Partnership and SuperTruck Initiative
Ken Howden, director, 21st Century Truck Partnership, DOE Office of Vehicle Technology

Reflections on GHG Phase I and Considerations for GHG Phase II
John Wall, vice president and chief technical officer, Cummins

Navistar Fuel Economy and Emissions
Greg Fadler, director of Performance Integration, Navistar

Lessons Learned from FE/GHG Phase 1 Regulations and Ways to Incorporate the Most Likely Future Technologies into FE/GHG Phase 2 Regulations
David Kayes, executive engineer, Compliance and Regulatory Affairs Department, Daimler Trucks North America

Overview of ICCT Heavy-Duty Vehicle Research Activities
Nicholas Lutsey, program director, International Council on Clean Transportation

Roadmap Findings and Technology Availability: Implications for GHG/Fuel Economy Regulations
Bill van Amburg, senior vice president, CALSTART

NAS Phase II Study of Medium-Duty and Heavy-Duty Truck Fuel Consumption
Timothy Blubaugh, executive vice president, Truck and Engine Manufacturers Association (EMA)

Strategic Outlook of North American Medium and Heavy Commercial Truck Powertrain Market
Sandeep Kar, global director, Commercial Vehicle Research, Frost & Sullivan

COMMITTEE SUBGROUP MEETING, MAY 9, 2013, WASHINGTON, D.C.

Closed sessions only

COMMITTEE MEETING 2, JUNE 20-21, 2013, WASHINGTON, D.C.

SwRI Work in Support of NHTSA and Its Relevance to NAS Study
Thomas Reinhart, Southwest Research Institute

Heavy-Duty Greenhouse Gas from a Full-Line Manufacturer's Perspective
Ken McAlinden, supervisor, Advanced Heavy-Duty Regulations, Ford Motor Co.

An Integrated, Heavy-Duty Vehicle Approach to GHG Regulation
Sam McLaughlin, external research manager, North American Region—Volvo Group Truck Technology

Fleet Operations and Fuel Consumption
Mike Roeth and Paul Menig

Question and Answer Session
Matthew Spears, Angela Cullen, Houshun Zhang, Byoungho Lee, and Prashanth Gururaja, EPA/Office of Transportation and Air Quality
OEM Experience with GHG Phase 1 and Recommendations for Phase 2
Michael Christianson, Daimler Trucks North America
Use of Vehicle Modeling in Engine Development at Cummins
Wayne Eckerle, vice president, Research and Technology Integration, Cummins
Gary Salemme, director, Advanced Engineering Systems Integration, Cummins
Engine Models and Maps Being Used in Truck Simulation
Nigel Clark, University of West Virginia

COMMITTEE MEETING 3, JULY 31 TO AUGUST 1, 2013, SACRAMENTO, CALIFORNIA

CalHEAT Technology Roadmap
Fred Silver and Tom Brotherton, CalSTART
California Efforts to Reduce Greenhouse Gas Emissions from Heavy-Duty Vehicles
Dan Sperling, member, California Air Resources Board; professor, Civil Engineering and Environmental Science and Policy, University of California at Davis
Natural Gas-Fueled Engines
Tim Frazier, director, Engineering, Cummins-Westport
National Academy of Science Panel on Heavy Duty GHG/CAFE Discussion: General Motors Comments
Mark Allen, director, Global Energy, Mass, and Aerodynamics, GM; Barbara Kiss, manager, Vehicle Efficiency and Energy Policy, GM
Natural Gas Station Infrastructure
Spencer Richley, Policy and Regulatory Associate, Clean Energy
Transforming Transportation: The Air Quality Need for Zero Emission Technologies
Matt Miyasato, deputy director, South Coast Air Quality Management District
Alternative and Renewable Fuel and Vehicle Technology Program
Janea Scott, commissioner, California Energy Commission
Legislation in California: Past and Present
Henry Stern, principal consultant, Senator Fran Pavley, Energy and Environmental Policy, California State Senate
Transmission Perspectives for the Phase 2 GHG Rule
Mihai Dorobontu, director, Technology Planning and Government Affairs, Eaton Vehicle Group

COMMITTEE MEETING, SEPTEMBER 25, 2013, WASHINGTON, D.C.

Closed sessions only

COMMITTEE MEETING 4, NOVEMBER 21, 2013, IRVINE, CALIFORNIA

Test Methods for Truck Tire Rolling Resistance and Reducing Fuel Consumption of Medium-Duty and Heavy-Duty Vehicles
Stan Lew, Industry Standards and Government Regulations, Michelin North America, Inc.
Demonstrated Efficiency of the achatesPOWER OP2S Engine
David M. Johnson, president and CEO, achatesPOWER
Transmission Technology and Fuel Consumption
Michael Howenstein, executive director, Strategic Controls, and Deborah Gordon, executive director, Business Planning and Program Management, Allison Transmission, Inc.
Presentation
Jay Spears, director, Standards and Regulation, and Curtis Decker, manager, Product Development, Continental Tire

COMMITTEE MEETING, JANUARY 30-31, 2014, WASHINGTON, D.C.

Closed sessions only

D

Acronyms and Abbreviations

AES	automatic engine shutdown
ALHT	automated long-haul truck
APU	auxiliary power unit
ATA	American Trucking Associations
ATIS	automatic tire inflation system
ATRI	American Transportation Research Institute
BEES	Board on Energy and Environmental Systems
BEV	battery electric vehicle
BTL	biomass to liquids
BTU	British thermal unit
CAFE	Corporate Average Fuel Economy
CalHEAT	California Hybrid, Efficient and Advanced Truck Research Center
CARB	California Air Resources Board
CFD	computational fluid dynamics
CH_4	methane
CI	compression ignition
CNG	compressed natural gas
CO	carbon monoxide
CO_2	carbon dioxide
DEF	diesel emission fluid
DME	dimethyl ether
DOE	U.S. Department of Energy
DOT	U.S. Department of Transportation
DPF	diesel particulate filter
EGR	exhaust gas recirculation
EIA	Energy Information Administration
EISA	Energy Independence and Security Act
EPA	U.S. Environmental Protection Agency
EPACT	Energy Policy Act of 2005
FC	fuel consumption
FCV	fuel-cell vehicle
FHWA	Federal Highway Administration
FMCSA	Federal Motor Carrier Safety Administration
FMEP	friction mean effective pressure
FT	Fischer-Tropsch (process)
FTA	Federal Transit Administration
FTP	Federal Test Procedure
GCVW	gross combination vehicle weight
GCVWR	gross combination vehicle weight rating
GDI	gasoline direct injection
GEM	GHG Emissions Model
GHG	greenhouse gas
GTI	Gasoline Technology Institute
GTL	gas to liquid
GVW	gross vehicle weight
GVWR	gross vehicle weight rating
HCCI	homogeneous-charge compression ignition
HEDGE	high-efficiency dilute gasoline engine
HEV	hybrid electric vehicle
HP	high pressure
HPDI	high-pressure direct injection
HVAC	heating, ventilation, and air conditioning
ISO	International Organization for Standardization
LCV	longer combination vehicle
LDV	light-duty vehicle
LED	light-emitting diode
LNG	liquefied natural gas
LRR	low rolling resistance
LSFC	load-specific fuel consumption
LTC	low-temperature combustion
MHDVs	medium- and heavy-duty vehicles
MTBE	methyl tertiary butyl ether
MY	model year

N_2O	nitrous oxide
NACFE	North American Council for Freight Efficiency
NG	natural gas
NHTSA	National Highway Traffic Safety Administration
NO_x	oxides of nitrogen
NPC	National Petroleum Council
NRC	National Research Council
NREL	National Renewable Energy Laboratory
OEM	original equipment manufacturer
OMB	Office of Management and Budget
ORNL	Oak Ridge National Laboratory
PCCI	premixed charge compression ignition
PCP	peak cylinder pressure
PFI	port fuel injection
PHEV	plug-in hybrid electric vehicle
PM	particulate matter
PNNL	Pacific Northwest National Laboratory
PTO	power take-off
R&D	research and development
RCCI	reactivity-controlled compression ignition
RFS	Renewable Fuels Standard
RFS2	Renewable Fuel Standard (2)
RIA	Regulatory Impact Analysis
RMA	Rubber Manufacturers Association
SAE	Society of Automotive Engineers
SCR	selective catalytic reduction
SET	supplemental emissions test
SI	spark-ignition
SUT	single unit truck
SwRI	Southwest Research Institute
TPMS	tire pressure monitoring (and maintenance) system
TPS	tire pressure systems
TWC	three-way catalyst
VIUS	Vehicle Inventory and Use Survey
VMT	vehicle-miles traveled
WBST	wide-base single tire
ZEV	zero emission vehicle

E

Glossary

NO_x or mononitrogen oxides: Refers to nitric oxide (NO) and nitrogen dioxide (NO_2), which are compounds formed during combustion.

Phase I Rule (or Phase I Regulations): Regulation jointly promulgated by the U.S. Environmental Protection Agency and the National Highway Traffic Safety Administration on September 15, 2011, and published in the Federal Register (76 Fed. Reg. 5710 et seq.).

Phase One Committee: Committee of the National Research Council, formally the Committee to Assess Fuel Economy Technologies for Medium- and Heavy-Duty Vehicles, which held its first meeting in December 2008 and delivered its final report, *Technologies and Approaches to Reducing the Fuel Consumption of Medium- and Heavy-Duty Vehicles*, in March 2010.

Phase One Report: Report issued in 2010 by the National Research Council and published by the National Academies Press entitled *Technologies and Approaches to Reducing the Fuel Consumption of Medium- and Heavy-Duty Vehicles.*

volume limited: For trucks involved in goods movement, "volume limited" refers to the situation in which the density of the freight is such that cargo volume is maximized without reaching the combined gross vehicle weight limit (typically 80,000 pounds for a Class 8 vehicle).

well to wheels: Specification for the envelope or boundary of an analysis defined to include and/or quantify the following: the extraction of the resource (e.g., petroleum or natural gas), the ultimate combustion of the fuel in the vehicle, and all steps in between to include refining, conversion, delivery, and so forth.

weight limited: For trucks involved in goods movement, "weight limited" refers to a situation in which the freight is sufficiently gravimetrically dense (as in the case of beverages, for instance) that the combined gross vehicle weight limit is reached (typically 80,000 pounds for a Class 8 vehicle).